Springer Theses

Recognizing Outstanding Ph.D. Research

Aims and Scope

The series "Springer Theses" brings together a selection of the very best Ph.D. theses from around the world and across the physical sciences. Nominated and endorsed by two recognized specialists, each published volume has been selected for its scientific excellence and the high impact of its contents for the pertinent field of research. For greater accessibility to non-specialists, the published versions include an extended introduction, as well as a foreword by the student's supervisor explaining the special relevance of the work for the field. As a whole, the series will provide a valuable resource both for newcomers to the research fields described, and for other scientists seeking detailed background information on special questions. Finally, it provides an accredited documentation of the valuable contributions made by today's younger generation of scientists.

Theses are accepted into the series by invited nomination only and must fulfill all of the following criteria

- They must be written in good English.
- The topic should fall within the confines of Chemistry, Physics, Earth Sciences, Engineering and related interdisciplinary fields such as Materials, Nanoscience, Chemical Engineering, Complex Systems and Biophysics.
- The work reported in the thesis must represent a significant scientific advance.
- If the thesis includes previously published material, permission to reproduce this must be gained from the respective copyright holder.
- They must have been examined and passed during the 12 months prior to nomination.
- Each thesis should include a foreword by the supervisor outlining the significance of its content.
- The theses should have a clearly defined structure including an introduction accessible to scientists not expert in that particular field.

More information about this series at http://www.springer.com/series/8790

Xiao-Sheng Zhang

Micro/Nano Integrated Fabrication Technology and Its Applications in Microenergy Harvesting

Doctoral Thesis accepted by
Peking University, Beijing, China

Author
Dr. Xiao-Sheng Zhang
Institute of Microelectronics
Peking University
Beijing,
China

and

Microsystems Laboratory
École Polytechnique Fédérale de Lausanne
Lausanne
Switzerland

and

Institute of Industrial Science
The University of Tokyo
Tokyo
Japan

Supervisor
Prof. Hai-Xia Zhang
Institute of Microelectronics
Peking University
Beijing
China

ISSN 2190-5053 ISSN 2190-5061 (electronic)
Springer Theses
ISBN 978-3-662-48814-0 ISBN 978-3-662-48816-4 (eBook)
DOI 10.1007/978-3-662-48816-4

Library of Congress Control Number: 2015955387

Springer Heidelberg New York Dordrecht London

Printed on acid-free paper

Springer-Verlag GmbH Berlin Heidelberg is part of Springer Science+Business Media
(www.springer.com)

Parts of this thesis have been published in the following journal articles:

1. Xiao-Sheng Zhang, M. D. Han, R. X. Wang, F. Y. Zhu, Z. H. Li, W. Wang, H. X. Zhang. Frequency-multiplication high-output triboelectric nanogenerator for sustainably powering biomedical microsystems. *Nano Letters*, 2013, *13*, 1168–1172.

2. Xiao-Sheng Zhang, F. Y. Zhu, M. D. Han, X. M. Sun, X. H. Peng, H. X. Zhang. Self-cleaning poly(dimethylsiloxane) film with functional micro/nano hierarchical structures. *Langmuir*, 2013, *29*, 10769–10775.

3. Xiao-Sheng Zhang, M. D. Han, B. Meng, H. X. Zhang. High performance triboelectric nanogenerators based on large-scale mass-production fabrication technologies. *Nano Energy*, 2015, *11*, 304–322. (Review Article)

4. Xiao-Sheng Zhang, M. D. Han, R. X. Wang, B. Meng, F. Y. Zhu, X. M. Sun, W. Hu, W. Wang, Z. H. Li, H. X. Zhang. High-performance triboelectric nanogenerator with enhanced energy density based on single-step fluorocarbon plasma treatment. *Nano Energy*, 2014, *4*, 123–131.

5. H. Zhang*, Xiao-Sheng Zhang*, X. Cheng*, Y. Liu, M. Han, X. Xue, S. Wang, F. Yang, A.S. Smitha, H. Zhang, Z. Xu. A flexible and implantable piezoelectric generator harvesting energy from the pulsation of ascending aorta: In vitro and in vivo studies. *Nano Energy*, 2015, *12*, 296–304. (*Co-First Author)

6. M. D. Han*, Xiao-Sheng Zhang*, B. Meng, W. Liu, W. Tang, X. M. Sun, H. X. Zhang. r-Shaped hybrid nanogenerator with enhanced piezoelectricity. *ACS Nano*, 2013, *7*, 8554–8560. (*Co-First Author)

7. Nicolas J. Peter*, Xiao-Sheng Zhang*, S. G. Chu, F. Y. Zhu, H. Seidel, H. X. Zhang. Tunable wetting behavior of nanostructured poly(dimethylsiloxane) by plasma combination treatments. *Applied Physics Letters*, 2012, *101*, 221601. (*Co-First Author)

8. Xiao-Sheng Zhang, B. Meng, F. Y. Zhu, W. Tang, H. X. Zhang. Switchable wetting & flexible SiC thin film with nanostructures for microfluidic surface-enhanced Raman scattering sensors. *Sensors and Actuators A: Physical*, 2014, *208*, 166–173.

9. Xiao-Sheng Zhang, Z. M. Su, M. D. Han, B. Meng, F. Y. Zhu, H. X. Zhang. Fabrication and characterization of the functional parylene-C film with micro/nano hierarchical structures. *Microelectronic Engineering*, 2015, *141*, 72–80.

10. Xiao-Sheng Zhang, Q. L. Di, F. Y. Zhu, G. Y. Sun, H. X. Zhang. Superhydrophobic micro/nano dual-scale structures. *Journal of Nanoscience and Nanotechnology*, 2013, *13*, 1539–1542.

11. Xiao-Sheng Zhang, F. Y. Zhu, G. Y. Sun, H. X. Zhang. Fabrication and characterization of squama-shape micro/nano multi-scale silicon material. *Science in China Series E*, 2012, *55*, 3395–3400.

12. Xiao-Sheng Zhang, Q. L. Di, F. Y. Zhu, G. Y. Sun, H. X. Zhang. Wideband anti-reflective micro/nano dual-scale structures: fabrication and optical properties. *Micro & Nano Letters*, 2011, *6*, 947–950.

Supervisor's Foreword

In the past decades, micro/nano science and technologies have grown very fast, and have already become one of the most significant disciplines supporting the rapid development of human society. In the long history of research and development, scientists have learned from nature, both from the macro-world and micro-world, and from the earth and planets and molecules and atoms. They all seem completely different but are actually tightly correlated with each other. In general, the micro-world is the footstone and building block, exploring it and revealing the secrets behind are of significance to reasonably explain the phenomena in the real macro-world.

To investigate the micro-world, we must solve several problems; the first is, can we see it? The invention of the scanning electron microscope in the beginning of the past century and its further development and commercialization in the following decades opened our "eyes" to this magic small world. The second question is how to make it? Due to the development of microfabrication technology, including photolithography, wet and dry etching, atomic deposition, and so on, the attractive vision of realizing the artificial micro-world has become true. Based on these efforts, scientists have moved forward to a deeper level of the micro-world, i.e., nanoscale, molecular or atoms, many new findings and techniques have been emerging in the past 30 years. Among this, an interdiscipline at the interface of microscale and nanoscale, i.e., micro/nano hierarchical structures, has attracted much attention resulting from possessing the advantages from both scales.

In fact, micro/nano hierarchical structures, known as micro/nano dual-scale structures, are widely distributed in the natural world and show many unique properties. For example, the self-cleaning property of lotus leaf results from the micro/nano hierarchical structures onto the surface, which consist of mastoid-shaped microstructures and villus-like nanostructures. Besides, the super adhesion of gecko's toes is attributed to the existence of micro/nano hierarchical structures. Therefore, people wish to imitate natural micro/nano hierarchical structures and strengthen their unique properties by fabricating well-designed structures. Although several techniques have been developed to fabricate micro/nano hierarchical structures, most of them suffer from disadvantages such as high cost, low production rate, noncompatibility with standard CMOS process, and so on.

Thus, exploring a novel micro/nano integrated fabrication approach has already become a research hot spot, which is essential for the development of micro/nano science and technology. In this thesis, a new cost-efficient, mass-fabrication, and universal technique for micro/nano integrated fabrication is proposed. The fundamental basis of this novel technique comes from the deep reactive-ion etching (DRIE) process. First, an improved DRIE process was developed to realize a maskless wafer-level fabrication technique that can be used to form Si-based micro/nano hierarchical structures. Second, a post-DRIE process was proposed as a plasma treatment process to modify the substrate to realize the ultralow-surface-energy surfaces.

Actually, the improved DRIE process and the post-DRIE process can be merged as a single-step DRIE process by simply optimizing the process recipe of the inductively coupled plasma (ICP) etcher. Then, based on the ultralow-surface-energy silicon substrate, a novel micro/nano integrated fabrication technique for flexible materials was developed using a single-step replication process. Consequently, a universal micro/nano integrated fabrication technology was successfully developed, which can be used to realize well-designed micro/nano hierarchical structures onto both silicon substrate and flexible materials.

Finally, in order to show the benefits of the micro/nano integrated fabrication technology presented here, this thesis also introduced it into the microenergy field and proposed several high-performance triboelectric nanogenerators (TENG). By electric measurement, theoretical analysis, finite element simulation and applications, the fabricated TENGs were demonstrated to be a robust micropower source, and were successfully utilized to power portable electronics and biomedical microsystems. This work opens a new chapter for applying micro/nano hierarchical structures for developing renewable micropower source.

This is a start for using micro-nano dual structure to make high-performance energy harvester which Dr. Xiaosheng Zhang has tried his best to discover and go as deep as as he can during his Ph.D. study period. It is exciting to see that it attracts more and more attention these days due to the emergence of portable electronics and micro systems.

Beijing, China
September 2015

Prof. Hai-Xia Zhang

Abstract

Micro/nano hierarchical structures in the natural and artificial worlds show attractive potential in many application fields, such as photoelectric conversion device, online detection of cells and biomaterials, self-cleaning material and super adhesion surface, due to their abundant unique properties. Several techniques have been developed to fabricate micro/nano hierarchical structures, which can be classified as Top-Down and Bottom-Up methods. However, conventional techniques suffer from some disadvantages, such as the limitation of minimum lithography scale, mask, low production rate, small process area, and so on, which hinder the rapid development and wider application of micro/nano hierarchical structures. Therefore, realizing a controllable, maskless, wafer-level, cost-efficient, and universal micro/nano integrated fabrication technology is significant and essential for the development of the micro/nano field.

Through experimental and theoretical studies, this thesis presents a systematic approach to fabricate wafer-level micro/nano hierarchical structures based on conventional microfabrication techniques. This fabrication approach has been demonstrated to be simple, reliable, compatible, mass-production, and low-cost, and has been used to realize silicon-based materials and flexible materials. The fabricated samples show several attractive properties such as super-hydrophobicity, wide-band anti-reflectance, and surface-enhanced Raman scattering. Moreover, this novel micro/nano integrated fabrication technology is introduced into the microenergy field and several high-performance triboelectric nanogenerators are fabricated and studied.

The research work in this thesis can be summarized as follows:

First, this thesis presents a single-step maskless nanoforest fabrication technique based on an improved deep reactive-ion etching (DRIE) process by the investigation and optimization of key process parameters and working range via a great many independent and comparative experiments. Subsequently, through experimental exploration and mechanism investigation, as well as characterization analysis, a wafer-level Si-based micro/nano integrated fabrication technique is realized by a combination of the above-improved DRIE process and the conventional microfabrication process. In the meantime, the multiscale interaction effect during the Si-based micro/nano hierarchical fabrication was investigated. The fabricated Si-based samples possess remarkable wide-band anti-reflectance and stable super-hydrophobicity, showing the attractive potential application future in photoelectric conversion device and micro/nano fluidics.

Second, the above Si-based micro/nano integrated fabrication technique is extended for flexible polymeric materials, and we propose a single-step fabrication technique to realize polymeric micro/nano hierarchical structures by replication process. The quantitative relation between key process parameters and replication precision has been figured out. The new phenomenon and fundamental rule of multiscale interaction effect during polymeric micro/nano hierarchical fabrication were also studied using the finite element simulation. The chemical modification of fluorocarbon plasma treatment for reducing the surface energy has been studied by density functional theory. After the above systematical investigation, we successfully fabricated micro/nano hierarchical structures on two common polymeric materials (i.e., PDMS and parylene-C), which show the outstanding super-hydrophobicity and surface-enhanced Raman scattering property.

Third, we employed this micro/nano integrated fabrication technology into the microenergy field to realize a high-performance sandwich-shaped triboelectric nanogenerator (TENG). The working principle of this novel device was studied by finite element simulation. The frequency response and the effect of structural size were investigated by detailed electric measurement. The ability to sustainably power micro device and system by this sandwich-shaped TENG has been demonstrated by successfully driving five parallel-connected LEDs.

Finally, this thesis further introduces this micro/nano integrated fabrication technology to TENG, and presents three universal methods to enhance the output performance of TENG. These three methods include single-step fluorocarbon plasma treatment, surface geometric optimization, and hybrid energy harvesting. The density functional theory (DFT) was employed to analyze the chemical modification mechanism of this fluorocarbon plasma treatment, modeling for the first time the energy required for electron transfer for different friction materials at molecular level based on first-principle calculations. The reliability, stability, and wide applicability of the fluorocarbon plasma treatment have been studied from several aspects, such as plasma treatment methods, plasma gas, substrate, continuous working ability, long-term test, etc. The enhancement of surface geometric optimization has been demonstrated by comparison among flat surface, nanostructured surface, microstructured surface, and micro/nano hierarchical structures. An r-shaped hybrid nanogenerator with piezoelectricity and triboelectricity has been fabricated based on the enhancement method of hybrid energy harvesting. Using the above high-performance flexible TENG, this thesis also presents some innovative applications in self-powered sensors, commercial electronics, and biomedical microsystems.

In summary, this thesis presents a universal mass-production micro/nano integrated fabrication technology, which can be used to realize micro/nano hierarchical structures on Si-based materials and flexible polymeric materials. This fabrication technology has been systematically investigated using experimental measurements, mechanism analyses, theoretical simulations, and so on. Three common materials (i.e., silicon, PDMS, and parylene-C) with micro/nano hierarchical structures have been successfully fabricated, which also show several attractive properties. Furthermore, this thesis introduces this fabrication technology in microenergy system, and proposes several high-performance nanogenerators, whose practical applications have also been studied in commercial electronic devices and biomedical microsystems.

Keywords Micro/nano hierarchical structures · Silicon · Flexible polymeric material · Nanogenerator · Microenergy source

Acknowledgments

My heart is full of emotions while writing this part. Looking back at the four years in Yanyuan, time has flown like a fleeting show. During my Ph.D. study period, there were many memories, happy, and sad, all of which are the most valuable treasures of my life. Just before the end of my Ph.D. life, I would like to express my heartfelt gratitude to all the people who have ever cultivated me, taught me, and helped me, and for those admirable teachers, those amiable and respectable seniors, and those kind and cute brothers and sisters. Thank you!

First of all, I would like to thank my supervisor Prof. Hai-Xia Zhang for her kind advice, strong support, and warm encouragement during my time at Peking University. Professor Zhang is not only an advisor for my research but also a mentor in my life. She once joked, "I 'picked' Xiaosheng into my pocket". "Pick" actually means especially concentrated and focused on doing things. When I was a child, I liked picking shells and pebbles along the river with other children. I always hoped to pick up the most beautiful one. I do not expect to become the most beautiful and gorgeous one you have "picked", but I do hope that with hard work, sweat, and achievements, I can become a heavy and fruitful shell. This Ph.D. thesis is the most comprehensive summary for my research work and it is also dedicated to Prof. Zhang, expressing my sincerest thanks!

Secondly, I would like to thank Profs. Zhi-Hong Li, Wen-Gang Wu, and Wei Wang. They are three respectable guides in my academic career. When I encountered bottlenecks and problems during my research, Prof. Li always gave me constructive comments. Professor Wu gave me a lot of tips and norms for how to summarize the research work and write English articles. Professor Wang always answered my questions patiently regarding experimental difficulties and failures. I still remember that I borrowed the oscilloscope from Prof. Li, Rhodamine reagent from Prof. Wu, PDMS solution from Prof. Wang, etc. These are the footstones during my growth, and they are also a part of my happy memories.

In addition, I would like to thank Profs. Yu-Feng Jin, Gui-Zhen Yan, Yi-Long Hao, Da-Cheng Zhang, Xiao-Mei Yu, Jin-Wen Zhang, Jing Chen, and Zhen-Chuan

Yang for their guidance and help, which will be my precious wealth in my future academic career.

Thank you Prof. Xin Zhao from Nankai University for giving me guidance on the project. Thank you Prof. Qing Chen from Peking University, Prof. Quan-Shui Zheng from Tsinghua University, Prof. Ya-Pu Zhao from Mechanics Institute of Chinese Academy of Sciences, Prof. Hong-Bo Sun from Jilin University, Prof. Wei-Zheng Yuan from Northwestern Polytechnical University for your guidance and help during the experiments and test analyses. Thank you Prof. Chih-Ming Ho from UCLA for giving me help and encouragement in my study. Thank you Profs. Qing Chen and Yan-Yi Huang from Peking University and Prof. Tian-Ling Ren from Tsinghua University for their precious comments and suggestions on my Ph.D. thesis.

I also thank Profs. Xiao-Bo Yang and Jing-Fu Bao, who supervised my research when I pursued my master's degree in the University of Electronic Science and Technology of China. Their careful cultivation laid a good foundation for my scientific research, and their warm words will benefit me in my whole life.

Thank you Mr. Zhi-Bo Ma from Northwestern Polytechnical University for giving me guidance and help in the experiment process. Thank you Dr. Guang-Yi Sun from Nankai University for your guidance and help in the black silicon processing experiments and English article writing.

Thank you Dong-Ming Fang, Xiu-Han Li and Zhe Chen for the care and help that always encouraged me to move forward. Thank you to the current and former members in Alice's Wonderlab, who fought side by side with me. Thank you to all my partners in room 329. Thank you Tim, Nico and Philip, I will never forget those happy times we spent together!

I also thank the editors and staff from Springer for helping me to prepare the paperwork for publishing my Ph.D. thesis, especially Wayne Hu, Ivy Gong and Fermine Shaly.

Finally, I thank my family with the sincerest feelings. Thank you to my parents, brother and sister-in-law, sister and brother-in-law, and two cute nieces for giving me love and happiness! I also thank my beautiful wife Jing Su. Without her love and strong support, I might never be able to make it.

Finally, I would like to say thanks truthfully to all the people who have encouraged, supported, and helped me in my life.

This Ph.D. thesis is supported by the National Natural Science Foundation of China (Grant nos. 61176103, 91023045 and 91323304), the National Hi-Tech Research and Development Program of China ("863" Project) (Grant no. 2013AA041102), and the Beijing Natural Science Foundation.

Contents

Abbreviations

ADF	Amsterdam Density Functional
CA	Contact Angle
CNT	Carbon Nanotube
DFT	Density Functional Theory
DRIE	Deep Reactive-Ion Etching
FEA	Finite Element Analysis
FTIR	Fourier Transform Infrared Spectrum
GGA	Generalized Gradient Approximation
IC	Integrated Circuit
ICP	Inductively Coupled Plasma
IoT	Internet of Things
LCD	Liquid Crystal Display
LED	Light Emitting Diode
LPCVD	Low Pressure Chemical Vapor Deposition
MEA	Microneedle Electrode Array
MEMS	Micro-Electro-Mechanical Systems
MNDS	Micro/Nano Dual-scale Structures
MNHS	Micro/Nano Hierarchical Structures
MNHS-G	Groove-shaped Micro/Nano Hierarchical Structures
MNHS-P	Pyramid-shaped Micro/Nano Hierarchical Structures
NG	Nanogenerator
P3HT	Poly(3-hexylthiophene)
PAA	Porous Anodic Alumina
PBS	Phosphate Buffered Saline
PC	Polycarbonate
PDMS	Poly(dimethylsiloxane)
RA	Rolling Angle
RIE	Reactive-Ion Etching

RSA	Ratio of Surface Area
SEM	Scanning Electron Microscope
SERS	Surface-Enhanced Raman Scattering
TENG	Triboelectric Nanogenerator
ZORA	Zero-Order-Regular Approximation

Chapter 1
Introduction

Abstract This chapter mainly reviews the development road map of micro-/nanointegrated fabrication technology as well as the previous research work of micro-/nanohierarchical structures. Consequently, the motivation, purpose, and innovative contributions of this thesis are briefly summarized.

As the development of science and technology step by step, the objective environment is explored gradually by human beings, and the scale of research subject becomes smaller and smaller from meter (m), centimeter (cm), millimeter (mm), to micrometer (μm), nanometer (nm), and picometer (pm), as shown in Fig. 1.1. The diameter of human's hair is around a few tens of or a hundred micrometers (i.e., 10^{-6} m), and this size is named as micrometer scale. When the size is reduced further by three orders of magnitude (i.e., 10^{-9} m), it is named as nanometer scale and most of the bacteria are at this level. As the fundamental footstone constructing the real world, the tiny subjects at micro- and nanolevels have attracted much attention, which are believed to be the key to explain the essential phenomena and problems of macroscale subjects.

Since the 1950s, people's understanding on microscale world has moved forward by a big step as the rapid development of microfabrication and microscopy technology. In the meantime, since the 1980s, the technology of nanoscale synthesis and characterization has promoted our knowledge of nanoscale world from the preliminary stage to the advanced stage. However, the significant research branch between microscale and nanoscale, which is named micro-/nanodual-scale structure (i.e., micro-/nanohierarchical structure), is still lack of sufficient investigation. There are plenty of various micro-/nanohierarchical structures not only in nature but also in the artificial world, and all of them show the unique and attractive properties. Therefore, the research on micro-/nanohierarchical structures and their applications in micro-/nanosystems is always an essential topic worldwide [1–8], and this thesis also focuses on this field.

X.-S. Zhang, *Micro/Nano Integrated Fabrication Technology and Its Applications in Microenergy Harvesting*, Springer Theses, DOI 10.1007/978-3-662-48816-4_1

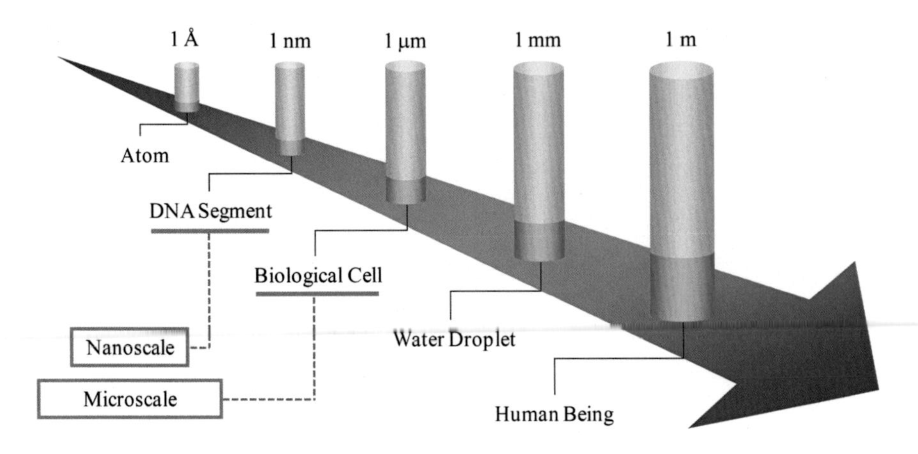

Fig. 1.1 Illustration of microscale and nanoscale

1.1 Micro-/Nanohierarchical Structures in Nature

The micro-/nanohierarchical structures widely exist in nature, and all of them possess the promising features. For example, when the rain droplets fall onto the surface of lotus leaves, they bound for several times and then slide immediately. As a famous phenomenon of super-hydrophobicity, even a tiny water droplet cannot stay atop a tilted surface of lotus leave. After a long-time and deep study, this unique property of lotus leaf is revealed by using the scanning electron microscope technique, as shown in Fig. 1.2.

As shown in Fig. 1.2, the surface of lotus leaf is constructed by the combination of microstructures and nanostructures, which are micro-/nanohierarchical structures. The microstructures are mastoid-shaped arrays with high density, and every microstructure is made of villus-like nanostructures. Thus, this kind of micro-/nanohierarchical structures sharply reduces the liquid–solid interface area resulting from enlarging the surface roughness. Then, more air can be trapped by this hierarchical structure and induces the super-hydrophobic feature, which is well known as self-cleaning property too. These micro-/nanohierarchical structures widely exist in nature, not only the lotus leaf but also many other plants [9–12]. There are also plenty of micro-/nanohierarchical structures in the animal world besides the plant world [13–17]. Let us take gecko for example, it can walk and even run on vertical walls due to the remarkable super-adhesion of its foot-fingers, and the secret of gecko's feet is the high-density micro-/nanohierarchical structures. The surface of a gecko foot-finger consists of hundreds of thousands of micropillars, and each micropillar is made of hundreds of nanopillars [13], which makes it possible to tightly adhere to the sidewall surface.

The above description clearly shows that there are various micro-/nanohierarchical structures in nature, and all of them possess attractive properties, such as super-hydrophobicity, anti-reflectance, and super-adhesion. These unique

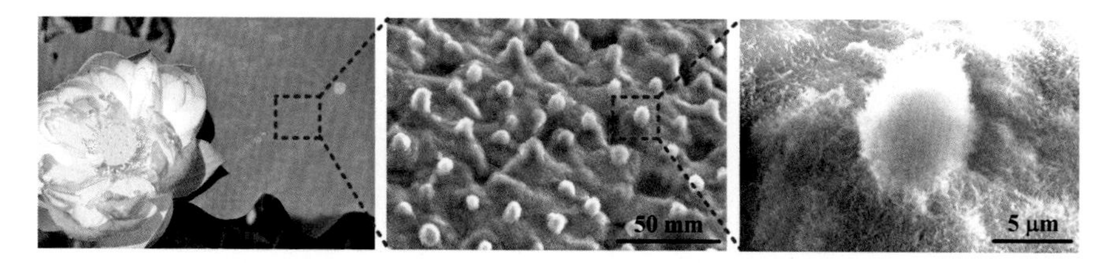

Fig. 1.2 Scanning electron microscope (SEM) images of lotus leaf, which consist of mastoid-shaped microstructures and villus-like nanostructures

properties have attracted much attention because they can be used to promote our life and benefit the society. The super-hydrophobicity can be used to realize the self-cleaning surface and then reduce the water consumption. The anti-reflectance can be utilized to enhance the performance of solar cell and give a positive response to the worldwide energy crisis. And the super-adhesion makes it possible to realize human's dream of walking on the sidewall. Therefore, the study on micro-/nanohierarchical fabrication technology is very important, which not only significantly contributes to the development of micro-/nanoscience but also greatly benefits our daily life.

1.2 Artificial Micro-/Nanohierarchical Structures

As mentioned in Sect. 1.1 above, micro-/nanohierarchical structures have attracted so much attention and have already become one of hot research spots due to their attractive features. The fabrication technology is the footstone and the fundamental point for the research on micro-/nanohierarchical structures. Several techniques have been developed to fabricate micro-/nanohierarchical structures, which can be summarized as two main methods, i.e., bottom-up and top-down. Each of them consists of several different processes.

1.2.1 The Progress of Bottom-Up Method

Bottom-up method can be simply defined as using small-size materials to construct big-size materials based on some kinds of "growing" processes, such as polymerization, synthesis, deposition, and crystallization. Additionally, the mentioned "small-size" material is a relative concept compared with the mentioned "big-size" materials. For example, several small-size molecules accumulate and combine to be a big-size molecule with a specific shape; or a small-size polymeric material is gradually deposited to be a big-size bulk material.

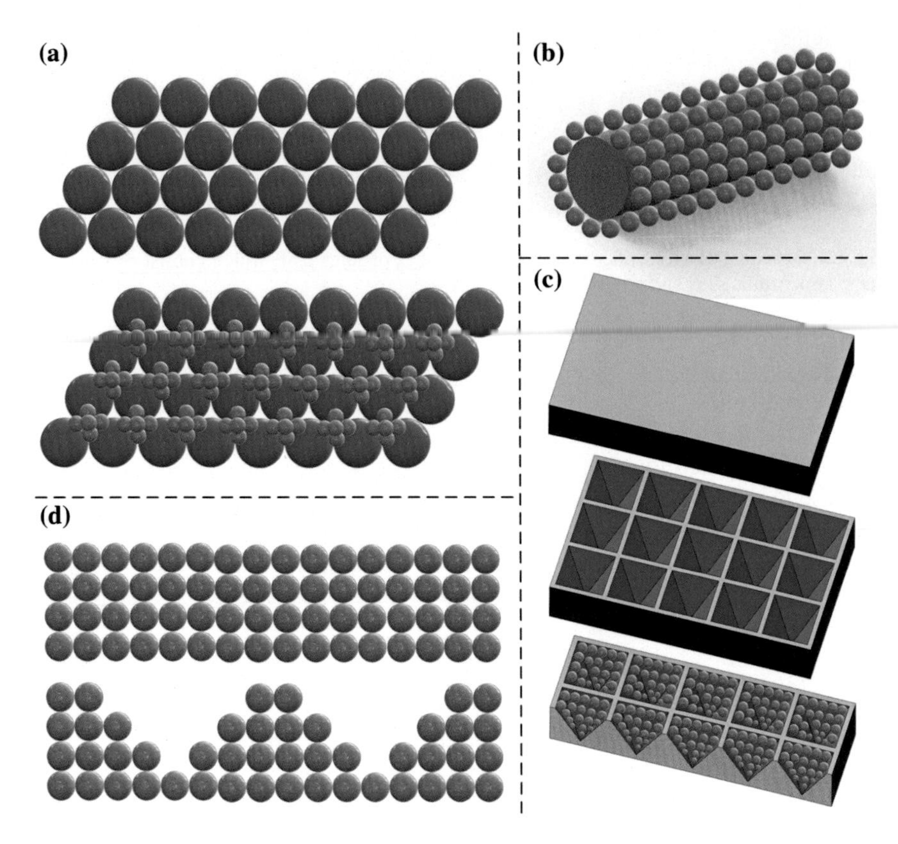

Fig. 1.3 Schematic views of four types of bottom-up method to realize micro-/nanohierarchical structures

The reported bottom-up methods can be summarized as four types, including directly growing, nanostructure growing atop microstructure, microfabrication combined with nanostructure growing, and nanostructure growing followed by microfabrication. Figure 1.3 illustrates four types of bottom-up method, and the detailed explanations are given as follows.

(i) **Directly Growing**

The directly growing process means the technique to fabricate both microstructures and nanostructures simultaneously by only using the growing method. In Ref. [18], the group led by *Hiroaki Imai* from Keio University reported a directly growing process to synthesis titanate (TiO_2) nanosheets with hierarchical structures, which consist of microstructures (spherical particles or hollow fibers) made of high-density nanosheets. Obviously, this single-step directly growing process is very simple. However, the speed of TiO_2 growing is very low and it is hard to control the growing of TiO_2. In Ref. [19], *Hongta Yang* from Chung Hsing University,

Peng Jiang from University of Florida, and their colleagues reported biomimetic hierarchical arrays based on a two-step self-assembly process. First, microscale SiO_2 balls were self-assembled on a substrate, and subsequently nanoscale SiO_2 balls were self-assembled atop the microball arrays. Eventually, the superhydrophobic hierarchical arrays made of self-assembled microballs and nanoballs were fabricated. The directly growing process is a two-step fabrication technique, i.e., fabricating microstructures followed by fabricating nanostructures, which is time-consuming.

(ii) **Nanostructure Growing Atop Microstructure**

The second bottom-up method is nanostructure growing atop microstructures, which can be simply described as fabricating nanostructures on the existing microstructures. Here, three publications are selected to explain the above process. In Ref. [20], the group led by *Lei Zhai* from University of Central Florida represented a bottom-up assembly process to fabricate poly(3-hexylthiophene) (i.e., P3HT) hierarchical structures by using carbon nanotubes (CNTs) as supporting structure. The length and density of P3HT hierarchical structures can be controlled by changing the weight ratio of P3HT to CNT. In Ref. [21], the group led by *Gregory N. Parsons* from North Carolina State University reported ZnO hierarchical structures produced by using sequential hydrothermal crystal synthesis and thin-film atomic layer deposition. In Ref. [22], the group led by *Huilan Su* from Shanghai Jiaotong University successfully grew high-density TiO_2 nanofibers atop $CaCO_3$ microfibers using a bio-inspired bottom-up assembly solution technique. In summary, the key point of the above method is fabricating nanostructures based on a bottom-up process on the existing microstructures.

(iii) **Microfabrication Combined With Nanostructure Growing**

The third bottom-up method to realize micro-/nanohierarchical structures is microfabrication combined with nanostructure growing, which can be divided into two steps, including microstructure fabrication and subsequent nanostructure growing. Obviously, the main difference between the second method (ii) and the third method (iii) is whether microstructures already exist. Here, five papers are selected to illustrate this method in detail.

In Ref. [23], the group led by *Jae-Min Myoung* from Yonsei University reported a biomimetic hierarchical ZnO structure with super-hydrophobic and anti-reflective properties. The fabrication process contains two steps, i.e., firstly fabricating microcraters by using anisotropic wet etching and then synthesizing ZnO nanorods onto the microcraters by using metal-organic-chemical-vapor deposition. In Ref. [24], the group led by *Qian Liu* from National Center for nanoscience and Technology, China, reported a BiOCl hierarchical structure fabricated by combining wet-etching (i.e., microfabrication) process with liquid-phase crystal growth (i.e., nanostructure growing) process. In Ref. [25], the group from Korea Institute of Science and Technology and Seoul National University reported 3-D hierarchical wrinkled micropillars on poly(dimethylsiloxane) (PDMS). Micropillars were firstly replicated from a silicon mold, and then, a diamond-like amorphous carbon layer was deposited by using radio frequency-chemical vapor deposition. In Ref. [26], the group led by *Xiaoniu Yang* from Changchun Institute of Applied Chemistry, Chinese Academy of Sciences, reported a process to realize hierarchical structures on poly(3-butylthiophene) (P3BT) substrate. A casting process was firstly employed to fabricate microstructures on P3BT substrate, and then, a controlled solvent vapor treatment was used to realize nanoscale fiber-like lamellae. In Ref. [27], the group led by *Junjie Li* and *Changzhi Gu* from Institute of Physics, Chinese Academy of Sciences, reported a process combining induction coupling plasma (ICP) etching and hot filament chemical deposition (HFCVD) to fabricated flower-like graphene hierarchical structures.

In summary, the third bottom-up method is not a pure bottom-up method, which combines a top-down method (i.e., microfabrication) and a bottom-up method (i.e., nanostructure growing) together.

(iv) **Nanostructure Growing Followed By Microfabrication**

The fourth bottom-up method to fabricate micro-/nanohierarchical structures is nanostructure growing followed by microfabrication. The fourth bottom-up method is similar to the third one, and both of them contain two steps (i.e., microfabrication and nanostructure growing) but with different sequences. Generally, the fourth bottom-up method can be described as follows, i.e., firstly nanostructures are formed by using growing method, and secondly parts of nanostructures are removed by using microfabrication, and finally micro-/nanohierarchical structures are formed. For example, the group led *Tatsuya Okubo* from The University of Tokyo reported a combined approach to fabricated hierarchical silica films [28]. A surfactant-directed self-assembly process was firstly used to prepare mesoporous silica films, and then reactive-ion etching (RIE) process was employed to generate vertically aligned pores on the above films.

In summary, all of these bottom-up methods shown in Fig. 1.3 possess a common key point of growing small-size structures (i.e., nanostructure growing). Although there are several techniques to realize nanostructure growing, most of them suffer the disadvantage of low growth speed. Additionally, the reliability and controllability of this bottom-up method are affected easily by experimental parameters and environmental factors. Therefore, it is difficult to use this bottom-up method for mass production of micro-/nanohierarchical structures.

1.2.2 The Progress of Top-Down Method

The top-down method comes from the mass-production technology of microelectronics, such as photolithography and etching processes. Based on these mass-fabrication techniques, people can reduce the dimension of big-size materials and realize small-size structures. Top-down method is widely used in microfabrication field to form well-designed microstructures and already becomes a perfect technical system that is suitable for various materials and can be employed to realize plenty

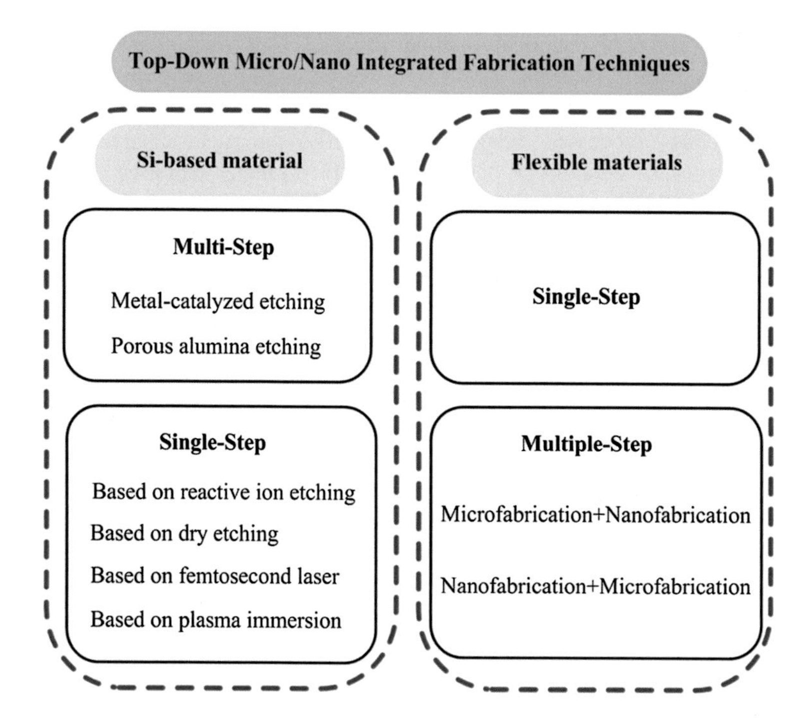

Fig. 1.4 Summary of top-down micro-/nanointegrated fabrication technology

of functional structures. In this technical system of top-down fabrication method, the most significant techniques can be summarized as follows, i.e., top-down micro-/nanointegrated fabrication techniques for Si-based material and flexible materials, as shown in Fig. 1.4. As is well known, due to the attractive properties, silicon is the perfect material for semiconductor industry as well as the most widely used material in microfabrication field. However, the fragile feature and biomedical incompatibility hinder the application of silicon in other fields. Thus, flexible materials, especially biocompatible ones, have attracted much attention and are considered as promising materials to overcome drawbacks of Si-based material.

Regardless of what kinds of materials used, top-down methods can be classified into two types, including multiple-step process and single-step process. In the following main text, top-down methods for both Si-based material and flexible materials are explained in detail.

(i) Top-Down Method For Silicon-Based Material

Multiple-step Processes

The multiple-step top-down fabrication methods for silicon mainly contain two types, including metal-catalyzed chemical etching process and porous anodic alumina (PAA) template-assisted wet-etching process.

For metal-catalyzed chemical etching process, four papers are selected to describe it in detail. In Refs. [29–33], researchers reported Si-based micro-/nanohierarchical structures fabricated by gold-, silver- and copper-catalyzed chemical etching processes, respectively. Briefly, the metal-catalyzed chemical etching process contains two main steps. First, metal particles are formed on a prepatterned (i.e., microstructured) silicon substrate and serve as catalysts. Second, the silicon substrate is immersed into the chemical solution, and then, the silicon will be etched under the catalysis effect of metal. According to the published work, normally nanostructures fabricated atop prepatterned silicon substrate are porous and rods with high density but very low aspect ratio. Furthermore, as a common disadvantage of wet-etching processes, this metal-catalyzed chemical etching process is not compatible with standard CMOS processes. Additionally, this process also brings a challenge to the silicon substrate, i.e., the pollution of heavy metal elements.

For PAA template-assisted wet-etching process, the group led by *Ronggui Yang* from University of Colorado at Boulder proposed it in Ref. [34]. In this work, the researchers firstly evaporated a thin aluminum (Al) film on the surface of prepatterned silicon substrate with microstructures, and then, the Al film was anodized in 0.3 M oxalic acid to form *in situ* PAA template. Subsequently, a thin gold layer was deposited to serve as catalyst, and then, the wafer was etched in the mixture solution of HF and H_2O_2. Finally, high-density nanorods formed atop microstructures and micro-/nanohierarchical structures formed on silicon substrate. Obviously, this process can be used to fabricate wafer-level samples directly. But, it contains five steps, including preparation of Al, anodization of Al, deposition of Au, chemical wet etching, and removal of Al and Au. Thus, it is complicated and incompatible to the standard CMOS processes.

In summary, multiple-step top-down micro-/nanointegrated fabrication techniques generally contain more than two steps even by using prepatterned substrates. As is well known, more fabrication steps make a process more complicated and then induce high cost and reduce reliability and controllability.

Single-step Processes

In order to overcome the above drawbacks of multiple-step processes, several single-step processes were also developed to fabricate Si-based micro-/nanohierarchical structures, which can be summarized as follows: RIE process, dry etching process, femtosecond laser etching process, and plasma immersion process.

In Ref. [35], the group led by *Yang Xia* from Lanzhou University proposed a plasma immersion process to fabricate nanoscale needle-like structures atop microstructures to form micro-/nanodual-scale textured silicon surface. In Ref. [36], the group led by *Junghoon Lee* from Seoul National University reported an isotropic dry etching process based on XeF_2 gas to fabricate nanograss atop micropillar arrays to form micro-/nanohierarchical structures on silicon substrate. In Ref. [37], the group led by *Feng Chen* from Xi'an Jiaotong University reported a femtosecond laser irradiation process to fabricate nanoporous atop microgrooves to form micro-/nanohierarchical structures. In Ref. [38], the group led by *JinsuYoo* from

Korea Institute of Energy Research proposed a RIE process to grow nanograss on the microstructured silicon substrate to realize micro-/nanohierarchical structures.

In summary, compared with multiple-step processes, the above single-step processes are very simple and cost-effective, which can be utilized to fabricate nanostructures directly on microstructures. However, they also suffer from some disadvantages, for example, it is hard to realize high-aspect-ratio structures and fabrication area of single process step is small.

(ii) **Top-Down Method For Flexible Materials**

Top-down methods for flexible materials are similar to those for silicon material, which can be also classified into two main kinds, including multiple-step processes and single-step processes.

Multiple-step Processes

The multiple-step top-down methods for flexible materials can be summarized as two types according to the sequence of microfabrication and nanofabrication. Fabricating microstructures and nanostructures sequentially is named "microfabrication + nanofabrication," while the opposite one is named "nanofabrication + microfabrication." In Ref. [39], the group led by Barbara Cortese from Universitàdel Salento reported a PDMS film with micro-/nanohierarchical structures. Micropillar arrays were fabricated on PDMS surface firstly by using replication process from silicon mold, and then, CF_4 plasma treatment was employed to modify PDMS surface further to form nanoscale patterns. This is a typical "microfabrication + nanofabrication" process. Reference [40] proposed the other process of "nanofabrication + microfabrication." First, an anodic aluminum oxide (AAO) membrane with periodic nanoporous was used as the template to realize nanostructures onto the polycarbonate (PC) film by embossing. And then, periodic microcylinders were formed on the nanostructured PC film by embossing from the mold. Finally, PDMS film was casted from the embossed PC film, and micro-/nanohierarchical structures formed atop PDMS film.

Single-step Processes

Besides the above multiple-step processes, several single-step processes were developed to fabricate flexible materials with micro-/nanohierarchical structures, which are simple and cost-efficient. Generally, the current single-step processes are based on the replication process (i.e., casting or molding) to replicate micro-/nanohierarchical structures directly from the mold. Reference [41] reported a PDMS micro-/nanohybrid surface replicated from a microstructured nanodimpled aluminum (MNA) master. By combining aluminum chemical oxidization and UV-photolithography, a MNA master with periodic microporous photoresist patterns atop the nanodimpled aluminum surface was realized. Then, the complimentary micro-/nanohierarchical structures were fabricated on top of PDMS surface by replicating from the MNA master. Similarly, Ref. [42] reported a PDMS artificial lotus leaf film by using single-step replication process, but the mold was prepared by using an under-exposed under-baked photolithography technique.

In summary, the single-step processes are based on the replication method to replicate micro-/nanohierarchical structures directly from the prepatterned mold. These methods are very simple but powerful to high efficiently produce flexible materials with micro-/nanohierarchical structures. However, looking into these processes deeply, we will find that surfactant is necessary for the traditional single-step methods, which is used to prevent the bonding between flexible materials and the mold and makes it easy to peel off flexible materials. Unfortunately, the surfactant covers the mold substrate, especially micro-/nanopatterns atop the mold, and thus, the replication precision is reduced. Furthermore, an additional process is necessary to remove the surfactant, which makes the whole process more complicated. Actually, it is very hard to remove it completely, and the mold surface is polluted. Therefore, a new fabrication technique is required to overcome these drawbacks.

1.3 The Development Progress of Microenergy Field

Due to the combination of microstructures and nanostructures, micro-/nanohierarchical structures show plenty of attractive properties and are widely used in many fields, including photoelectric conversion, biomedical detection, self-cleaning materials, and super-adhesion surface. In particular in new energy source field, it has attracted much attention, and several high-performance solar cells have already been developed based on the anti-reflectance of micro-/nanohierarchical structures. However, in microenergy field of powering miniature devices or even microsystems, it is still lack of knowledge to apply micro-/nanohierarchical structures. But according to the remarkable benefits from applying either microstructures or nanostructures in microenergy field, researchers believe that micro-/nanohierarchical structures have attractive potentials.

1.3.1 Microenergy Technology

The modern society are facing four challenges, including energy and food crises, and population and environmental problems. As the economy increases year by year, the worldwide energy consumption enlarges sharply. The traditional energy sources (i.e., fossil fuels), such as coal, oil and gas, are non-renewable and limited, which bring the serious environmental problems. Thus, several new energy sources, such as wind energy, hydropower, and solar energy, which are clean and renewable. Besides the above power sources, there is still an important energy form, i.e., microenergy.

Microenergy is defined as an energy source powering miniature devices and systems, and the feature size is at centimeter, micrometer, or even nanometer levels. Although the energy consumption all over the world has increased to be at TW level, the power consumption of electronic products decreases continuously

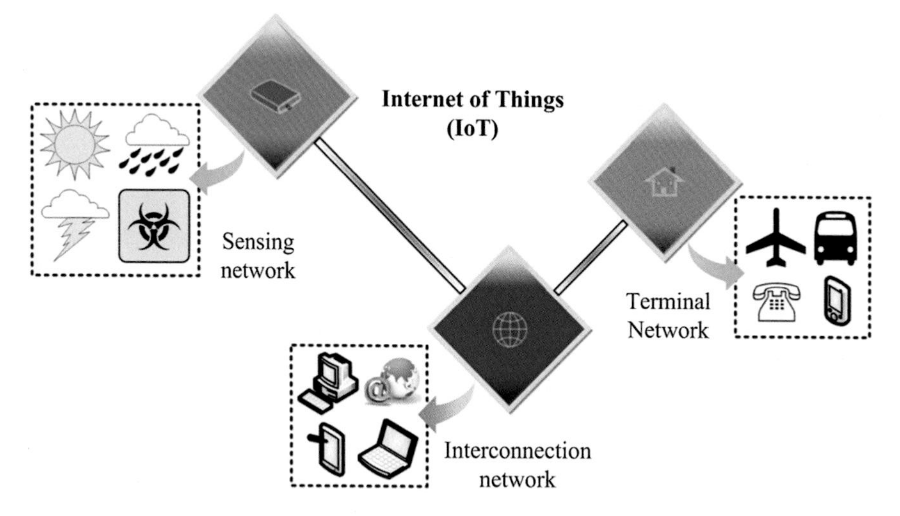

Fig. 1.5 Schematic view of the construction of the internet of things (*IoT*)

to be at mW or even μW levels [43]. However, the amount of electronic products increase quickly every year; thus, their total power consumption achieves a huge level. Therefore, developing cost-efficient microenergy sources is very important for the development of microelectronic devices and systems. Let us take the Internet of Things (IoT) for example as shown in Fig. 1.5.

As the next generation of hybrid smart network, IoT is considered as an essential support for the development of economy and society, which is playing an important role in our daily life. Generally, IoT system consists of three main parts, including sensing network, interconnection network, and terminal network. The various changes in the environment are collected and transferred to be electronic signals by sensing network. Then, these electronic signals are delivered to the interconnection network and are processed to form control signals. Finally, the control signals are transmitted to the terminal network, and the terminal electronic devices will respond to the environmental changes. Obviously, as the interface media between environment and client, the sensing network is the fundamental part of IoT. The sensing network is composed of plenty of sensors, which are widely distributed in the environment, especially in remote places. Therefore, how to sustainably power the sensing network has already become an urgent issue for the development of IoT, and the autonomous and self-powered concept were proposed. The microenergy sources harvesting energy from the living environment have been demonstrated as one of the most efficient approaches. Based on the piezoelectric effect, electromagnetic effect, and pyroelectric effect, etc., microenergy sources can accumulate the energy with various forms and transform it to electric power to drive miniature devices and systems. These microenergy sources are green, sustainable, and cost-efficient, more importantly, it is very easy to integrate these microenergy sources with other electronic products.

1.3.2 Nanogenerator

Since it was proposed at the beginning of twenty-first century, as a rapid developing green energy source, the nanogenerator has attracted much attention. Nanogenerator is a power source applying nanoscience and nanotechnology for harvesting energy for powering micro-/nanosystems, which was highlighted by Discovery magazine as one of 20 important inventions in nanoscience field [44].

The first nanogenerator was developed by *Z.L. Wang* from Georgia Institute of Technology in 2006, which can be used to harvest mechanical energy from the environment based on zinc oxide (ZnO) [45]. After nine-year development, including experimental exploration and theoretical analysis, material preparation and device fabrication, and various applications, nanogenerators have already become an independent research area [46, 47]. The reported nanogenerators can be classified into five types, including vertical nanowires, lateral nanowires, non-contact nanowires, nanofibers, and hybrid type [44]. When ZnO nanowires are fabricated in the vertical direction, it is named nanogenerators based on the vertical nanowires [48]. In contrast, when ZnO nanowires are prepared in the lateral direction, it is called nanogenerators based on lateral nanowires [49]. When an external force is applied, a voltage potential will be generated at the two ends of nanowires. For these two types above, it is easy to realize high-density ZnO nanowires. Thus, their output performances are relatively high.

A non-contact nanogenerator was also developed, in which ZnO nanowires were packaged and insulated by PMMA [50]. Two electrodes were placed in the opposite sides of PMMA. When an external force is applied to the device, the effective voltage potential will be generated in the force direction. Besides nanowires, piezoelectric fibers were also introduced to construct nanogenerator [51, 52], which made the device very sensitive to the force change. Additionally, in order to enhance the output performance, several hybrid nanogenerators were designed, which combined piezoelectric effect with other energy-harvesting mechanisms, such as solar energy [53].

During the past years, the output performance of nanogenerators increased gradually, and the voltage and the power density has achieved 10–50 V and 10–500 mW/cm^3, respectively, in 2011 [47]. Moreover, piezoelectric nanogenerators were widely utilized as sensors to detect the pressure change due to unique piezoelectric properties. Then, the self-powered active sensors were developed without external power source. However, the output performance of piezoelectric nanogenerator is still insufficient to supply microdevices and systems, which limits its wider applications. The main reason resulting in this disadvantage is the limited electric power generating by a single nanowire. Although several techniques were developed to fabricate nanowire arrays, the density of nanowire arrays is not sufficient. Additionally, the fragile feature of piezoelectric nanowire reduces the reliability and stability of nanogenerator, and it is hard to realize flexible nanogenerators. Thus, a novel nanogenerator based on new working principle was proposed to solve these problems and enhance the output performance.

1.3.3 Triboelectric Nanogenerator

In January 2012, a novel nanogenerator based on the combination of triboelectrification effect and electrostatic induction was developed by the group led by *Z.L. Wang* from Georgia Institute of Technology [54]. It is named triboelectric nanogenerator (TENG). TENG has attracted much attention during the past three years due to its attractive properties, including clean, cost-efficient, and high performance. For example, from 2012 to 2013, the power density of TENG increased sharply from 3.67 mW/m^2 to 313 W/m^2, as shown in Fig. 1.6.

The first TENG was simply made of two films of polyester and Kapton; thus, its output was not high and the voltage was 3.3 V. However, as the first work, this prototype is very important, which establishes the footstone of TENGs' development and opens a new chapter of microenergy field. Subsequently, TENG develops very fast in different aspects, including operation mechanisms, structural optimization, and novel materials. Basically, the development of TENG can be summarized as new structures, new materials, and new principles.

New Structures

As well known, larger roughness of triboelectric surfaces means larger effective triboelectric area, in other words, the triboelectrification effect is stronger. Then, more charges are generated on triboelectric surfaces, and the output performance of TENG is higher. Thus, nanostructures and microstructures were fabricated as triboelectric surfaces to enlarge the roughness, and the output of TENGs increased efficiently [55–57]. Furthermore, it was found that the separating speed

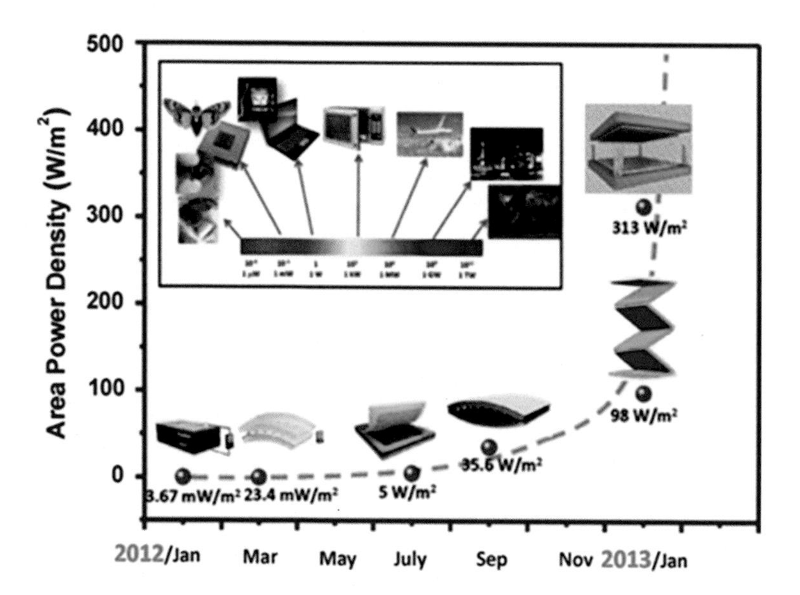

Fig. 1.6 The development road map of triboelectric nanogenerator from 2012 to 2013. Reproduced from Ref. [43]. Copyright 2013 American Chemical Society

of triboelectric pairs tightly affects the TENG output performance; in other words, the gap between triboelectric pairs is the key point when their mechanical properties are fixed. Therefore, an arch-shaped structure was introduced in TENG to realize the rapid separation and enhance the electric output [57]. The TENG based on the combination of microstructured surface and arch-shaped structure showed remarkable output performance with a voltage of 230 V [57].

New Materials

During the past three years, researchers have already tried many materials to enhance the output performance of TENG, and the key point behind materials is the triboelectric series. Triboelectric series is a ranking table according to the ability of acquiring/losing electrons during the triboelectric effect, as shown in Table 1.1 [58, 59]. The bigger ranking number means the stronger ability of acquiring electrons, while the smaller ranking number means the stronger ability of

Table 1.1 Triboelectric series of common materials

Triboelectric series: ranking table of the ability of acquiring/losing electrons					
No.	Material	No.	Material	No.	Material
1	Aniline–formol resin	17	Styrene–acrylonitrile copolymer	33	Polyacrilonitrile
2	Polyformaldehyde 1.3-1.4	18	Styrene–butadiene copolymer	34	Acrylonitrile–vinyl chloride
3	Etylcellulose	19	Wood	35	Polybisphenol carbonate
4	Polyamide 11	20	Hard rubber	36	Polychloroether
5	Polyamide 6-6	21	Acetate, Rayon	37	Polyvinylidine chloride (Saran)
6	Melanimeformol	22	Polymethyl methacrylate (Lucite)	38	Poly(2,6-dimethyl polyphenylene oxide)
7	Wool, knitted	23	Polyvinyl alcohol	39	Polystyrene
8	Silk, woven	24	Polyester (Dacron) (PET)	40	Polyethylene
9	Polyethylene glycol succinate	25	Polyisobutylene	41	Polypropylene
10	Cellulose	26	Polyurethane flexible sponge	42	Polydiphenyl propane carbonate
11	Cellulose acetate	27	Polyethylene terephthalate	43	Polyimide (Kapton)
12	Polyethylene glycol adipate	28	Polyvinyl butyral	44	Polyethylene terephtalate
13	Polydiallyl phthalate	29	Formo-phenolique, hardened	45	Polyvinyl chloride (PVC)
14	Cellulose (regenerated) sponge	30	Polychlorobutadience	46	Polytrifluorochloroethylene
15	Cotton, woven	31	Butadiene–acrylonitrile copolymer	47	Polytetrafluoroethylene (Teflon)
16	Polyurethane elastomer	32	Natural rubber		

losing electrons. When the ranking numbers of triboelectric pairs are very close, the relative ability of acquiring/losing charges is very close. Then, it is very hard for charge transfer, and thus, the amount of generated charges are small. In contrast, when the ranking number difference of triboelectric pairs is very large, the relative ability difference of acquiring/losing electrons is very large. In other words, one material is easy to acquire electrons, while the other is easy to lose electrons. Then, it is very easy for charge transfer at the interface, and the amount of generated charges are very big. Therefore, the larger the relative difference of triboelectric pairs in triboelectric series, the better the output performance of TENG.

New Principles
Finally, one of the most important directions of TENGs' development is exploring new working principles, including two aspects: triboelectric modes and hybrid TENG. In the early period, TENGs were designed as contact-mode devices (i.e., vertical electrification). Subsequently, sliding-mode TENGs were developed (i.e., lateral electrification), which is very easy to harvest various natural energies (i.e., wind, hydropower, etc.) by using the rotation method [60, 61]. Additionally, TENGs were integrated with other energy-harvesting forms, such as piezoelectric effect, electromagnetic effect, and biochemical energy, to form hybrid TENGs, which can be utilized to harvest multiple kinds of energy simultaneously [62, 63].

As the investigation was further carried out, the output performance of TENG was enhanced gradually. A novel sliding-mode TENG based on surface-nanostructured nylon and PTFE was reported on April, 2013, whose open-circuit voltage, short-circuit current density, and power density achieved 1300 V, 4.1 mA/m^2 and 5.3 W/m^2, respectively [64]. One year later, a rotary TENG was reported, whose output power and power density achieved 1.5 W and 19 mW/cm^2, respectively [65].

1.4 Research Purpose and Content

1.4.1 Research Purpose

In summary, both natural micro-/nanohierarchical structures and artificial ones show plenty of attractive properties, such as super-hydrophobicity, super-hydrophilicity, anti-reflectance, and super-adhesion. And these remarkable features are very useful for social development and people's daily life, and they have attractive potential in various fields, such as biomedical detection, energy harvesting, and optical communication.

In particular, in microenergy field, due to maximizing the surface-area-to-volume ratio, micro-/nanohierarchical structures can be used to enlarge the efficiency of energy conversion, which already becomes a promising approach to realize the high-performance microenergy source. Additionally, realizing the high-performance microenergy source requires a mass-production micro-/nanointegrated fabrication technology, which should be cost-efficient and suitable for most of common materials (i.e., universal).

Furthermore, as the connect point of microscale structures and nanoscale structures, micro-/nanohierarchical structures possess all advantages from both of them and also show several novel features. These novel features brought by the combination of microstructures and nanostructures are very important for people exploring and understanding micro-/nanoworlds.

Therefore, the research work on mass-production micro-/nanointegrated fabrication technology is of great significance for the development of micro-/nanoscience, and benefitting the human society.

As described in the above sections, the fabrication techniques of micro-/nanohierarchical structures can be classified into two methods, including bottom-up method and top-down method. Bottom-up method is usually used to fabricate nanostructures, which is high-precision and high-controllability but low-speed, high-cost and material-limited. In contrast, top-down method is normally used to fabricate microstructures, which is mass production and large size but limited by the minimal size of photolithography.

Thus, developing a novel micro-/nanointegrated fabrication technique, which should be simple, cost-efficient, universal and mass production, has already become the core scientific issue of micro-/nanoscience and technology. However, the interaction among multiphysics field, the complex structures of materials, and the complicated conversion of energy makes it very hard to investigate micro-/nanohierarchical structures deeply and systematically. Generally, the minimal size of photolithography, the size effect and the interaction of multiple scales have already become the critical challenges of micro-/nanointegrated fabrication technology. Therefore, the research purpose of this work is overcoming the above drawbacks and proposing a novel micro-/nanointegrated fabrication technology, which is based on the traditional microfabrication technology but free of the minimal size of photolithography. Moreover, this novel micro-/nanointegrated fabrication technology is introduced in microenergy field in order to successfully solve the key problems of nanogenerators and realize several high-performance flexible nanogenrators.

1.4.2 Research Content

In order to realize the above purpose, this research work was carried out from four aspects, as shown in Fig. 1.7. The research content of this work can be summarized as follows, including novel fabrication techniques, new materials preparation, novel features exploration, and new devices development.

The organization of this book is briefly introduced as follows.

Chapter 1 Introduction

This chapter summarizes the current progresses of micro-/nanohierarchical structures and fabrication techniques and then analyzes the development of microenergy, especially nanogenerators. Subsequently, the purpose and main content of this research work are given.

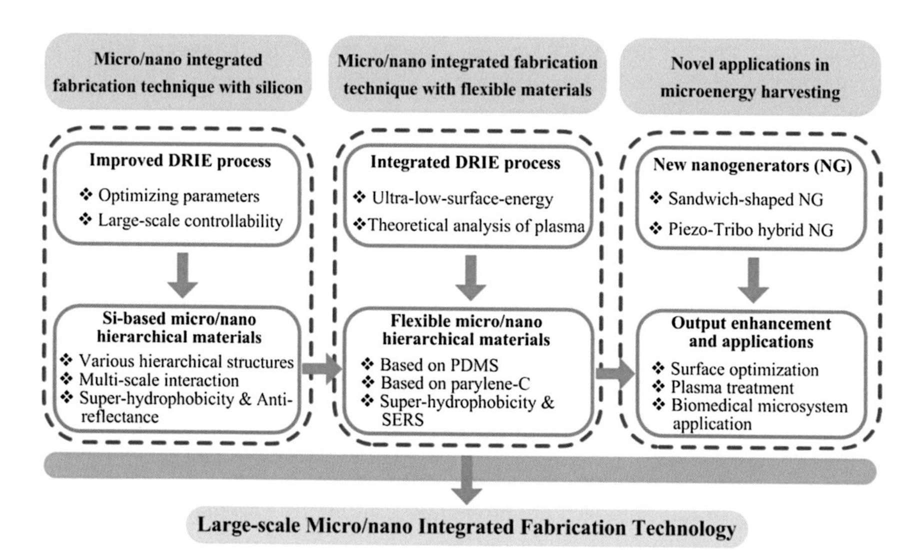

Fig. 1.7 Summary of research content of this book

Chapter 2 Micro/Nanointegrated Fabrication Technique for Silicon
A novel Si-based micro-/nanointegrated fabrication technique is developed based on the improved deep reactive-ion etching (DRIE) process. The key process parameters, working range, and optimized parameters are obtained. The interaction of multiple scales based on the silicon material is investigated. After the geometry design and the structural optimization, the optical property and the wettability of fabricated samples are studied, which shows super-hydrophobicity and wideband anti-reflectance.

Chapter 3 Micro-/Nanointegrated Fabrication Technique for Flexible Materials
Based on the combination of the above Si-based fabrication technique and the replication process of flexible material, a micro-/nanointegrated fabrication technique based on the flexible materials is proposed. The controllable fabrication of ultra-low-surface-energy materials and key parameters of replication process are investigated. In the meantime, the interaction of multiple scales based on flexible materials is first studied. Subsequently, both PDMS and parylene-C micro-/nanohierarchical structures are successfully fabricated. And then, the wettability of polymeric material is investigated at the molecular level based on the density functional theory (DFT) for the first time. The fabricated samples shows outstanding hydrophobicity and surface-enhanced Raman scattering property.

Chapter 4 Flexible Triboelectric Nanogenerators: Principle and Fabrication
The present micro-/nanointegrated fabrication technology is successfully introduced in nanogenerator field, and a novel flexible sandwich-shaped TENG was proposed. The working principle of this TENG is studied by finite element analysis (FEA) simulation. The electric property of fabricated TENG is

systematically investigated from several aspects, including frequency effect of external force, size effect, and surface profile. Additionally, as the first demonstration, TENG is successfully applied to drive commercial electronics and biomedical microsystems.

Chapter 5 Flexible Triboelectric Nanogenerators: Enhancement and Applications

Generally, two approaches are proposed to enhance the output performance of flexible TENGs. The first approach is single-step fluorocarbon plasma treatment based on the above micro-/nanointegrated fabrication technology. The DFT is employed for the first time to analyze the chemical modification mechanism of this fluorocarbon plasma treatment by calculating the energy required for electron transfer between different friction materials at the molecular level. Moreover, this enhancement approach is systematically studied from three aspects, including plasma gas, plasma generation, and different substrates. Additionally, the reliability and stability of this approach are proven by continuous working and long-term test. The second approach is based on the combination of different energy-harvesting mechanisms (i.e., hybrid TENG), and the piezoelectric-triboelectric nanogenerators is developed. Eventually, the applications of TENG in various fields, including biomedical microsystems, self-powered sensor network and portable electronics, are explored.

Chapter 6 Summary and Perspectives

The research work present in this book is summarized in this chapter, and the prospect is also pointed out.

References

1. R. Lakes, Materials with structural hierarchy. Nature **361**, 511–515 (1993)
2. V. Zorba, E. Stratakis, M. Barberoglou, E. Spanakis, P. Tzanetakis, S.H. Anastasiadis, C. Fotakis, Biomimetic artificial surfaces quantitatively reproduce the water repellency of a lotus leaf. Adv. Mater. **20**, 4049–4054 (2008)
3. H. Sai, K.W. Tan, K. Hur, E. Asenath-Smith, R. Hovden, Y. Jiang, M. Riccio, D.A. Muller, V. Elser, L.A. Estroff, S.M. Gruner, U. Wiesner, Hierarchical porous polymer scaffolds from block copolymers. Science **341**, 530–534 (2013)
4. R. Blossey, Self-cleaning surfaces-virtual realities. Nat. Mater. **2**, 301–306 (2003)
5. H. Cho, J. Kim, H. Park, J.W. Bang, M.S. Hyun, Y. Bae, L. Ha, D.Y. Kim, S.M. Kang, T.J. Park, S. Seo, M. Choi, K.Y. Suh, Replication of flexible polymer membranes with geometry-controllable nano-apertures via a hierarchical mould-based dewetting. Nat. Commun. **5**, 3137 (2014)
6. F. Xia, L. Jiang, Bio-inspired, smart, multiscale interfacial materials. Adv. Mater. **20**, 2842–2858 (2008)
7. W.R. Wei, M.L. Tsai, S.T. Ho, S.H. Tai, C.R. Ho, S.H. Tsai, C.W. Liu, R.J. Chung, J.H. He, Above-11 %-efficiency organic-inorganic hybrid solar cells with omnidirectional harvesting characteristics by employing hierarchical photon-trapping structures. Nano Lett. **13**, 3658–3663 (2013)
8. G.D. Bixler, B. Bhushan, Bioinspired rice leaf and butterfly wing surface structures combining sharkskin and lotus effects. Soft Matter. **8**, 11271–11284 (2012)

9. Y.Y. Yan, N. Gao, W. Barthlott, Mimicking natural superhydrophobic surfaces and grasping the wetting process: A review on recent progress in preparing superhydrophobic surfaces. Adv. Colloid Interface Sci. **169**, 80–105 (2011)
10. W. Barthlott, C. Neinhuis, Purity of the sacred lotus, or escape from contamination in biological surfaces. Planta **202**, 1–8 (1997)
11. Z. Guo, W. Liu, Biomimic from the superhydrophobic plant leaves in nature: Binary structure and unitary structure. Plant Sci. **172**, 1103–1112 (2007)
12. B. Bhushan, Y.C. Jung, K. Koch, Micro-, nano- and hierarchical structures for superhydrophobicity, self-cleaning and low adhesion. Philos. Trans. Roy. Soc. A. **367**, 1631–1672 (2009)
13. Y. Zhang, Y. Chen, L. Shi, J. Li, and Z. Guo, Recent progress of double-structural and functional materials with special wettability. J. Mater. Chem. **22**, 799–815 (2012)
14. B. Bhushan, Y.C. Jung, Natural and biomimetic artificial surfaces for superhydrophobicity, self-cleaning, low adhesion, and drag reduction. Prog. Mater. Sci. **56**, 1–108 (2011)
15. D. Byun, J. Hong, Saputra, J.H. Ko, Y.J. Lee, H.C. Park, B.K. Byun, J.R. Lukes, Wetting characteristics of insect wing surfaces. J. Bionic Eng. **6**, 63–70 (2009)
16. S.M. Lee, J. Üpping, A. Bielawny, and M. Knez, Structure-based color of natural petals discriminated by polymer replication. ACS Appl. Mater. Interfaces. **3**, 30–34 (2011)
17. K. Autumn, M. Sitti, Y.A. Liang, A.M. Peattie, W.R. Hansen, S. Sponberg, T.W. Kenny, R. Fearing, J.N. Israelachvili, R.J. Full, Evidence for van der Waals adhesion in gecko setae. *PNAS* **99**, 12252–12256 (2002)
18. Y. Takezawa, H. Imai, Bottom-up synthesis of titanate nanosheets with hierarchical structures and a high specific surface area, Small **2**, 390–393 (2006)
19. H. Yang, X. Dou, Y. Fang, P. Jiang, Self-assembled biomimetic superhydrophobic hierarchical arrays. J. Colloid Interface Sci. **405**, 51–57 (2013)
20. J. Liu, J. Zou, L. Zhai, Bottom-up assembly of poly(3-hexylthiophene)on carbon nanotubes: 2D building blocks fornanoscale circuits. Macromol. Rapid Commun. **30**, 1387–1391 (2009)
21. J.S. Na, B. Gong, G. Scarel, G.N. Parsons, Surface polarity shielding and hierarchical ZnO nano-architectures produced using sequential hydrothermal crystal synthesis and thin film atomic layer deposition. ACS Nano **3**, 3191–3199 (2009)
22. Q. Dong, H. Su, W. Cao, D. Zhang, Q. Guo, Y. Lai, Synthesis and characterizations of hierarchical biomorphic titania oxide by a bio-inspired bottom-up assembly solution technique. J. Solid State Chem. **180**, 949–955 (2007)
23. J. Xiong, S.N. Das, B. Shin, J.P. Kar, J.H. Choi, J.M. Myoung, Biomimetic hierarchical ZnO structure with superhydrophobic and antireflective properties. J. Colloid Interface Sci. **350**, 344–347 (2010)
24. Y. Tian, C.F. Guo, Y. Guo, Q. Wang, Q. Liu, BiOCl nanowire with hierarchical structure and its Raman features. Appl. Surf. Sci. **258**, 1949–1954 (2012)
25. Y. Rahmawan, K.R. Lee, M.W. Moon, K.Y. Suh, 3-D hierarchical wrinkled micro-pillars for anti-cells proliferation surface, in *6th IEEE Nanotechnology Materials and Devices Conference*, pp. 416–419, Jeju, Korea, 18–21 Oct 2011
26. G. Lu, L. Li, S. Li, Y. Qu, H. Tang, X. Yang, Constructing thin polythiophene film composed of aligned lamellae via controlled solvent vapor treatment. Langmuir. **25**, 3763–3768 (2009)
27. S. Tian, L. Li, W. Sun, X. Xia, D. Han, J. Li, C. Gu, Robust adhesion of flower-like few-layer graphene nanoclusters. Sci. Rep. **2**, 551 (2012)
28. K. Ijichi, A. Fukuoka, A. Shimojima, M. Sugiyama, T. Okubo, A combined top-down and bottom-up approach to fabricate silica films with bimodal porosity. Mater. Lett. **65**, 828–831 (2011)
29. Y. Xiu, L. Zhu, D.W. Hess, C.P. Wong, Hierarchical silicon etched structures for controlled hydrophobicity/superhydrophobicity. Nano Lett. **7**, 3388–3393 (2007)
30. F. Toor, H.M. Branz, M.R. Page, K.M. Jones, H.C. Yuan, Multi-scale surface texture to improve blue response of nanoporous black silicon solar cells, Appl. Phys. Lett. **99**, 103501 (2011)

31. X. Li, B.K.T ay, P. Miele, A. Brioude, D. Cornu, Fabrication of silicon pyramid/nanowire binary structure with superhydrophobicity. Appl. Surf. Sci. **255**, 7147–7152 (2009)
32. J.P. Lee, S. Choi, S. Park, Extremely superhydrophobic surfaces with micro- and nanostructures fabricated by copper catalytic etching. Langmuir. **27**, 809–814 (2011)
33. Y. He, C. Jiang, H. Yin, J. Chen, W. Yuan, Superhydrophobic silicon surfaces with micro-nano hierarchical structures *via* deep reactive ion etching and galvanic etching. J. Colloid Interface Sci. **364**, 219–229 (2011)
34. W. Wang, D. Li, M. Tian, Y.C. Lee, R. Yang, Wafer-scale fabrication of silicon nanowire arrays with controllable dimensions. Appl. Surf. Sci. **258**, 8649–8655 (2012)
35. J. Liu, B. Liu, S. Liu, Z. Shen, C. Li, Y. Xia, A simple method to produce dual-scale silicon surfaces for solar cells. Surf. Coat. Technol. **229**, 165–167 (2013)
36. Y. Kwon, N. Patankar, J. Choi, J. Lee, Design of surface hierarchy for extreme hydrophobicity. Langmuir **25**, 6129–6136 (2009)
37. D. Zhang, F. Chen, G. Fang, Q. Yang, D. Xie, G. Qiao, W. Li, J. Si, X. Hou, Wetting characteristics on hierarchical structures patterned by a femtosecond laser. J. Micromech. Microeng. **20**, 075029 (2010)
38. J. Yoo, G. Yu, J. Yi, Large-areamulticrystallinesiliconsolarcellfabricationusingreactiveion etching(RIE). Solar Energy Mater. Solar Cells **95**, 2–6 (2011)
39. B. Cortese, S. D'Amone, M. Manca, I. Viola, R. Cingolani, G. Gigli, Superhydrophobicity due to the hierarchical scale roughness of PDMS surfaces. Langmuir **24**, 2712–2718 (2008)
40. Y.H. Huang, J.T. Wu, S.Y. Yang, Direct fabricating patterns using stamping transfer process with PDMS mold of hydrophobic nanostructures on surface of micro-cavity. Microelectron. Eng. **88**, 849–854 (2011)
41. D.S. Kim, B.K. Lee, J. Yeo, M.J. Choi, W. Yang, T.H. Kwon, Fabrication of PDMS micro/nano hybrid surface for increasing hydrophobicity. Microelectron. Eng. **86**, 1375–1378 (2009)
42. Y. Yoon, D.W. Lee, J.H. Ahn, J. Sohn, J.B. Lee, One-step fabrication of optically transparent polydimethylsiloxane artificial lotus leaf film using under-exposed under-baked photoresist mold, in *25th IEEE International Conference on Micro Electro Mechanical Systems*, pp. 301–304, Paris, France, 29 Jan–2 Feb 2012
43. Z.L. Wang, Triboelectric nanogenerators as new energy technology for self-powered systems and as active mechanical and chemical sensors. ACS Nano **7**, 9533–9557 (2013)
44. Z.L. Wang, *Nanogenerators for Self-powered Devices and Systems* (Georgia Institute of Technology, Atlanta, 2011)
45. Z.L. Wang, J. Song, Piezoelectric nanogenerators based on zinc oxide nanowire arrays. Science **312**, 242–246 (2006)
46. Z.L. Wang, W. Wu, Piezotronics and piezo-phototronics-fundamentals and applications. Nat. Sci. Rev. **1**, 62–90 (2014)
47. Z.L. Wang, W. Wu, Nanotechnology-enabled energy harvesting for self-powered micro-/nanosystems. Angew. Chem. Int. Ed. **51**, 11700–11721 (2012)
48. X. Wang, J. Song, J. Liu, Z.L. Wang, Direct-current nanogenerator driven by ultrasonic waves. Science **316**, 102–105 (2007)
49. S. Xu, Y. Qin, C. Xu, Y. Wei, R. Yang, Z.L. Wang, Self-powered nanowire devices. Nat. Nanotechnol. **5**, 366–373 (2010)
50. R. Yang, Y. Qin, C. Li, G. Zhu, Z.L. Wang, Converting biomechanical energy into electricity by a muscle-movement-driven nanogenerator. Nano Lett. **9**, 1201–1205 (2009)
51. Y. Hu, Y. Zhang, C. Xu, G. Zhu, Z.L. Wang, High-output nanogenerator by rational unipolar assembly of conical nanowires and its application for driving a small liquid crystal display. Nano Lett. **10**, 5025–5031 (2010)
52. Y. Qin, X. Wang, Z.L. Wang, Microfibre nanowire hybrid structure for energy scavenging. Nature **451**, 809–813 (2008)
53. C. Chang, V.H. Tran, J. Wang, Y.K. Fuh, L. Lin, Direct-write piezoelectric polymeric nanogenerator with high energy conversion efficiency. Nano Lett. **10**, 726–731 (2010)

54. C. Xu, Z.L. Wang, Compact hybrid cell based on a convoluted nanowire structure for harvesting solar and mechanical energy. Adv. Mater. **23**, 873–877 (2011)
55. F.R. Fan, Z.Q. Tian, Z.L. Wang, Flexible triboelectric generator. Nano Energy **1**, 328–334 (2012)
56. G. Zhu, C. Pan, W. Guo, C.Y. Chen, Y. Zhou, R. Yu, Z.L. Wang, Triboelectric-generator-driven pulse electrodeposition for micropatterning. Nano Lett. **12**, 4960–4965 (2012)
57. F.R. Fan, L. Lin, G. Zhu, W. Wu, R. Zhang, Z.L. Wang, Transparent triboelectric nanogenerators and self-powered pressure sensors based on micropatterned plastic films. Nano Lett. **12**, 3109–3114 (2012)
58. S. Wang, L. Lin, Z.L. Wang, Nanoscale triboelectric-effect-enabled energy conversion for sustainably powering portable electronics. Nano Lett. **12**, 6339–6346 (2012)
59. A.F. Diaza, R.M. Felix-Navarro, A semi-quantitative tribo-electric series for polymeric materials: the influence of chemical structure and properties. J. Electrostat. **62**, 277–290 (2004)
60. P. Bai, G. Zhu, Y. Liu, J. Chen, Q. Jing, W. Yang, J. Ma, G. Zhang, Z.L. Wang, Cylindrical rotating triboelectric nanogenerator. ACS Nano **7**, 6361–6366 (2013)
61. L. Lin, S. Wang, Y. Xie, Q. Jing, S. Niu, Y. Hu, Z.L. Wang, Segmentally structured disk triboelectric nanogenerator for harvesting rotational mechanical energy. Nano Lett. **13**, 2916–2923 (2013)
62. Y. Yang, H. Zhang, Y. Liu, Z.H. Lin, S. Lee, Z. Lin, C.P. Wong, Z.L. Wang, Silicon-based hybrid energy cell for self-powered electrodegradation and personal electronics. ACS Nano **7**, 2808–2813 (2013)
63. Y. Yang, H. Zhang, J. Chen, S. Lee, T.C. Hou, Z.L. Wang, Simultaneously harvesting mechanical and chemicalenergies by a hybrid cell for self-powered biosensors and personal electronics. Energy Environ. Sci. **6**, 1744–1749 (2013)
64. S. Wang, L. Lin, Y. Xie, Q. Jing, S. Niu, Z.L. Wang, Sliding-triboelectric nanogenerators based on in-plane charge-separation mechanism. Nano Lett. **13**, 2226–2233 (2013)
65. G. Zhu, J. Chen, T. Zhang, Q. Jing, Z.L. Wang, Radial-arrayed rotary electrification for high performance triboelectric generator. Nat. Commun. **5**, 3426 (2014)

Chapter 2
Micro-/Nanointegrated Fabrication Technique for Silicon

Abstract Silicon is one of the most important materials for the development of electronics science and technology, which is the footstone of integrated circuits (IC), electronic communication systems, solar cells, micro-electro-mechanical systems (MEMS), etc. Since the 60s and 70s in the last century, the high-precision fabrication technology based on silicon has been developed rapidly. In the past five decades, Si-based microfabrication technology was developed from two aspects, including bulk processes and surface processes, which has already become a mature technical field. More importantly, bulk processes realize the dream of fabricating 3-D structures at the microscale level. Among all of the bulk processes, deep etching process is one of the most important techniques. Deep reactive-ion etching (DRIE) process is an essential deep etching process, and high-aspect-ratio structures are fabricated by using the alternation of etching steps and passivation steps. DRIE process was also called Bosch process due to the first developer of Bosch company. This chapter presents a micro-/nanointegrated fabrication technique for silicon based on an improved DRIE process, and several Si-based samples with attractive properties are demonstrated.

2.1 Nanoforest Fabrication Based on an Improved DRIE Process

2.1.1 Deep Reactive-Ion Etching (DRIE) Process

Generally, plasma etching processes can be classified into two main types, including reactive-ion etching (RIE) process and deep reactive-ion etching (DRIE) process. Both of these two processes utilize plasma to impact the substrate to realize the etching; however, their working principles are different. RIE process only contains the etching step, and there is only one radio frequency (RF) source to ionize

X.-S. Zhang, *Micro/Nano Integrated Fabrication Technology and Its Applications in Microenergy Harvesting*, Springer Theses, DOI 10.1007/978-3-662-48816-4_2

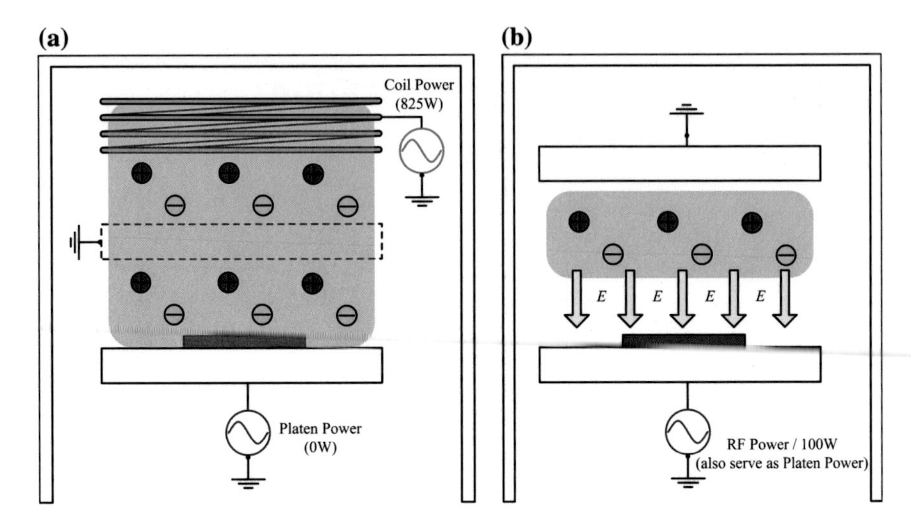

Fig. 2.1 The working principles of **a** deep reactive-ion etching (DRIE) process and **b** reactive-ion etching (RIE) process

the gas molecule and generate the plasma, and this RF source also serves as a platen power to accelerate the ions to impact the substrate. The working principle of RIE is shown in Fig. 2.1a. If neglect the direction influence from the RF source accelerating the plasma, the chemical reaction plays a dominant role in the etching process of plasma treatment on the substrate. And therefore, the etching rate in all directions approximately equals under the ideal condition, i.e., isotropic etching.

DRIE process is a highly anisotropic etching process, and it contains two steps, etching and passivation. Etching step utilizes high-concentration plasma ion to do physical impact and chemical etching, and the passivation step is using another plasma gas to treat the surface and form a polymeric layer by polymerization reactions. The etching step in DRIE process is basically consistent with the RIE etching mechanism, i.e., isotropic etching. However, in the standard DRIE process, etching steps and passivation steps proceed alternately. By using the protection of polymeric layer deposited by passivation steps to the vertical surface, the high-aspect-ratio structure can be realized step by step. And therefore, from the point of view of the final etching result, DRIE process is considered as a highly anisotropic etching process. To realize the above function, DRIE equipment contains tow RF sources. One is used to generate the plasma, and the other (i.e., platen power source) is used to control the internal electrical field. The working principle of DRIE is shown in Fig. 2.1b.

According to the above description, it is obvious that DRIE process has better controllability and wider application in the microfabrication field compared with RIE process, due to its two independent RF sources. Based on the method of generating plasma ions, DRIE process equipment has two types, inductively coupled

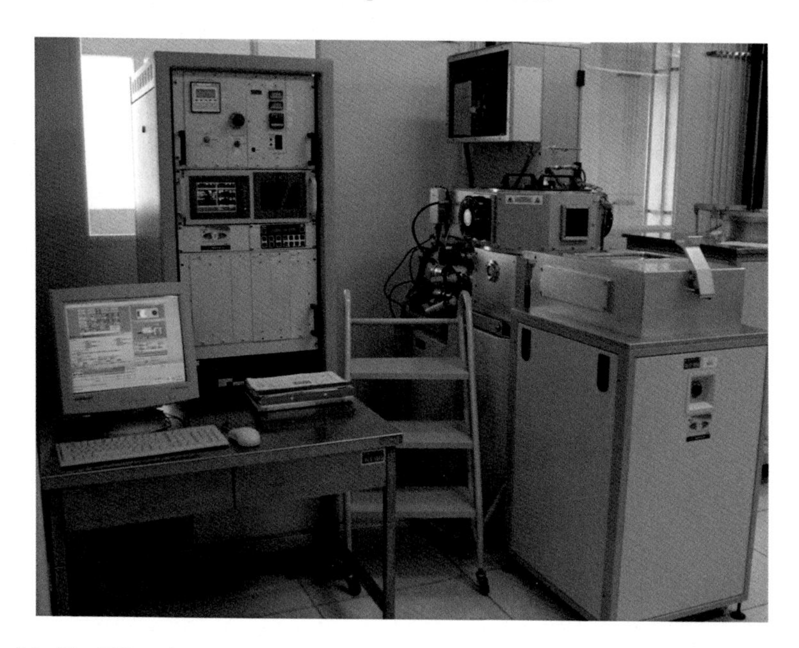

Fig. 2.2 The ICP equipment applied in the DRIE process

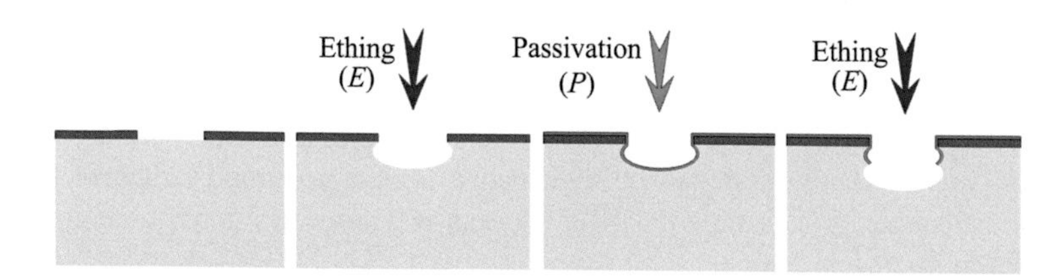

Fig. 2.3 The working principle of standard Bosch DRIE process

plasma (ICP) and capacitance-coupled plasma (CCP). The Multiplex ICP 48443 equipment produced by Surface Technology Systems (STS) Company is a typical one. And in this thesis, we used this equipment as shown in Fig. 2.2 to perform related experiments.

In the standard Bosch DRIE process, two types of gases are alternately used in the reaction chamber to realize the etching and passivation steps, respectively. These gases are formed high-dense plasma ions via glow discharge by RF power. The working principle of standard Bosch DRIE process is shown in Fig. 2.3.

The DRIE process contains complicated physical and chemical reactions. Basically, the etching plasma ions react with Si-based substrate and the formed gas is then released, while the passivation plasma ions form a polymer film on the surface of Si-based substrate. The chemical reactions are shown in Eqs. (2.1) and (2.2). As a common recipe of using SF_6 (etching gas) and C_4F_8 (passivation gas), the chemical reactions are described as follows.

$$SF_6\begin{cases} SF_6 \rightarrow S_xF_y^+ + S_xF_y + F^- \\ S_i + F \rightarrow S_iF_x \uparrow \\ S_xF_y^+ + S_iO_xF_y + e^- \rightarrow SO_xF_y \uparrow + S_iF_x \uparrow \\ (CH_3)_3[(CH_3)_2SiO]_nSi(CH_3)_3 + SF_6 \\ \rightarrow SO_xF_y \uparrow + SiF_x \uparrow + SiO_2 \end{cases} \tag{2.1}$$

$$C_4F_8\begin{cases} C_4F_8 \rightarrow CF_x^+ + CF_x + F^- \\ CF_x \rightarrow (CF_2)_n \end{cases} \tag{2.2}$$

During etching steps, complicated chemical reactions will happen between etching gas plasma ions and the substrate, and the etching rates in all directions are approximately same, which can be considered as isotropic etching step. However, due to the acceleration of platen power, the etching rate in the vertical direction is larger. During passivation steps, a polymeric thin film is formed via polymerization reactions on the substrate surface resulting in covering and protecting the substrate surface. However, due to the vertical directionality of etching steps, the polymeric film in the horizontal direction can be removed completely, and then, the substrate is further etched downwards, while the polymeric film on the vertical surface cannot be completely removed by etching steps, which protects the substrate in the vertical direction. Consequently, as the continuous alternation of etching steps and passivation steps, the high-dense high-aspect-ratio structure forms onto the substrate. The standard DRIE process is a core top-down technique of bulk silicon microfabrication.

2.1.2 *Nanostructuring by an Improved DRIE Process*

Based on the standard Bosch DRIE process, we proposed a maskless wafer-level nanofabrication technique based on an improved DRIE process [1–4]. By optimizing the process parameters of DRIE process, the polymer deposited during passivation steps cannot be removed completely. Thus, the residual polymeric nanoparticles serve as self-masks to protect the substrate during the etching steps, which results in high-dense high-aspect-ratio nanopillars. The schematic illustration of the formation of nanostructures by the improved DRIE process is shown in Fig. 2.4a, b shows SEM images of the growth procedure of nanostructures.

Basically, the optimization of DRIE process parameters is enhancing the passivation step or weakening the etching step. By a large amount of comparative experiments with one variable parameter, four key process parameters for the nanostructure formation were determined, i.e., gas flow, platen power, time ratio of etching and passivation (E/P), and total cycles. More importantly, different

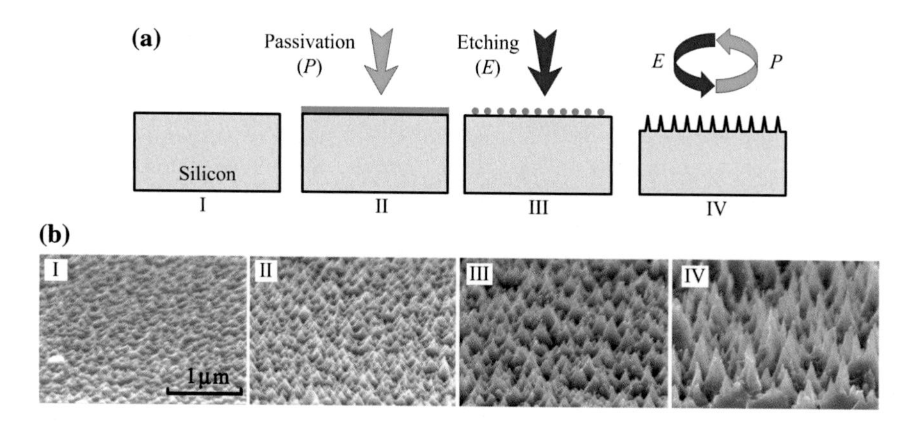

Fig. 2.4 Formation of nanostructures by an improved maskless DRIE process. **a** Schematic illustration. **b** SEM images

Table 2.1 The working ranges of DRIE process

Type	Gas flow (sccm)		Time ratio of E/P[a] (s:s)	Platen power (W)	Total cycles
	Etching	Passivation			
Normal etching	100	70	4/5	10	16
Hybrid etching	30	50	4/5	10	80
Etching stop	30	50	4/5	5	16
	30	50	4/5	6	40
Nanopillar forest	30	50	4/5	8	40

[a]E/P refers to etching/passivation

working ranges (i.e., fabrication results) of DRIE process were observed by the combinations of the above key parameters, as shown in Table 2.1.

Lots of comparative experiments show that the formation of nanostructures tightly correlates with the gas flow, which directly determines the working range of DRIE process. While other parameters, including total cycles, platen power, and E/P, have the significant influence on the profile of nanostructures. In summary, the gas flow controls the DRIE working range, while the other three parameters realize the controllable formation of nanopillar forest.

2.1.3 *Mechanism of Controllable Formation of Nanostructures*

Controllability is one of the key factors to assess whether a fabrication technique has an attractive application future. As mentioned in Sect. 2.1.2, the profile of nanostructures can be controlled by adjusting E/P, total cycles, and platen power.

As shown in Fig. 2.4b, although the fabricated nanostructures are non-uniform, the statistics data of their geometries are highly consistent in the large area. Thus, in order to better characterize and analyze the fabrication results, all the structural parameters are the average data in a unit area of 5 μm × 5 μm in the following discussion. According to the optical test and the wettability characterization, although the individual nanostructures are non-uniform, but the properties of fabricated samples are determined by the statistical data of nanopillars in a certain area, i.e., average structural data.

The profile of nanostructures can be defined by aspect ratio (*AR*) and density, where *AR* means the ratio of height (*h*) to diameter (*d*) at the half-height position, i.e., $AR = h/d$, and the density refers to the space (*s*) of adjacent nanopillars, as shown in Fig. 2.5. According to the comparative experiments, the density is extremely sensitive to the gas flow, and the optimal high-dense nanostructures (i.e., s = 100–400 nm) can only be obtained when the flows of etching gas and passivation gas are set as 30 sccm and 50 sccm, respectively. Besides, the quantitative relations between *AR* and the other three process parameters, i.e., E/P, total cycles, and platen power, are shown in Fig. 2.6.

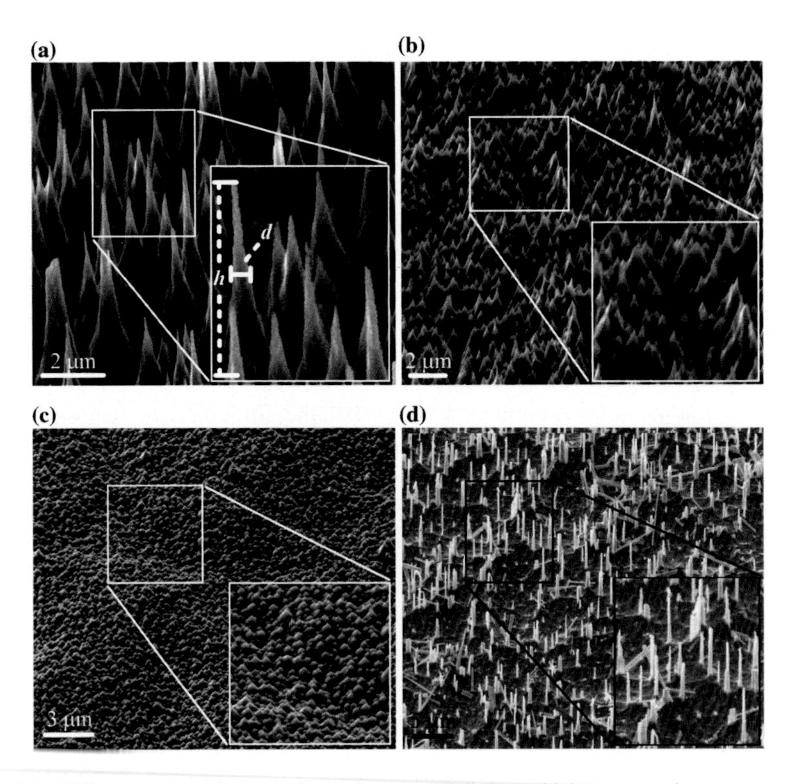

Fig. 2.5 Different types of nanoscale surfaces. **a** High-dense high-aspect-ratio nanostructures (space: 100–400 nm, *AR*: 8:1–13:1). **b** High-dense medium-aspect-ratio (space: 100–400 nm, *AR*: 3:1–7:1). **c** High-dense low-aspect-ratio (space: 100–400 nm, *AR*: 1:1–2:1). **d** Low-dense high-aspect-ratio (space: 1–10 μm, *AR*: 8:1–13:1)

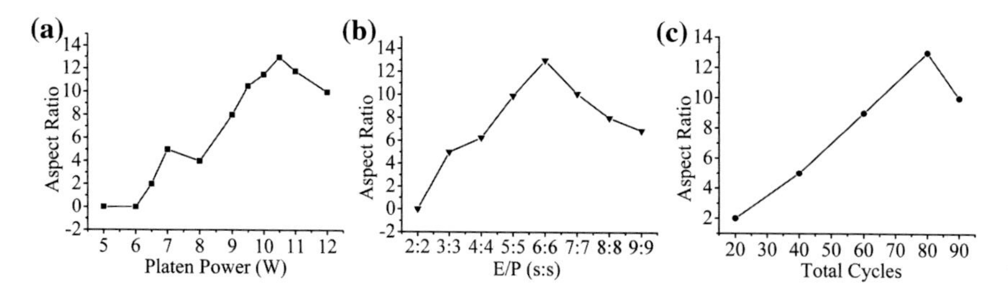

Fig. 2.6 The effects of key process parameters on the aspect ratio of nanostructures

Fig. 2.7 The fabricated 4-in. wafer-level sample

The curves in Fig. 2.6 reveal single-peak distribution, and in general, the curves tend to increase to a maximum value and then decrease from that point. The maximum *AR*s appear at platen power of 10.5 W, E/P of 6s:6s, and total cycles of 80. This trend is due to the overetching phenomenon. After *AR* reaches the maximum point, the further etching will reduce the height of nanopillar and then lower its *AR*. Based on this improved DRIE process, nanostructures with different profiles can be fabricated onto the substrate by quantitatively adjusting the process parameters. As shown in Fig. 2.5, four different nanostructures were realized: (a) high-dense high-aspect-ratio type, (b) high-dense medium-aspect-ratio type, (c) high-dense low-aspect-ratio type, and (d) low-dense high-aspect-ratio type. More importantly, this novel fabrication technique can be used to realize nanostructures onto the wafer-level substrate shown in Fig. 2.7, which shows the attractive ability of large-scale mass production.

2.2 Fabrication of Si-based Micro-/Nanohierarchical Structures

Based on the combination of traditional microfabrication techniques and the improved DRIE process for maskless nanofabrication, we can realize the Si-based micro-/nanohierarchical structures. As mentioned above, this new nanofabrication approach is maskless, simple, controllable, cost-efficient and mass production, compared with traditional nanofabrication methods.

2.2.1 Structural Design

Microstructures show various types and geometries corresponding to different application fields. However, the microstructure with periodic arrays is one of the most important types, which is essential in many fields. Besides, the micro-/nanostructures with periodic arrays is an effective method to test the reliability and homogeneity of this new micro-/nanointegrated fabrication technique. Here, in order to extend the application fields of this work, we selected two typical microstructures with inverted pyramid arrays and V-shaped-groove arrays, respectively, as shown in Fig. 2.8. The microstructure with inverted pyramid arrays is widely used in the optical field, and one typical application is serving as the light-trapping structure for high-performance solar cells. The microstructure with V-shaped-groove arrays is widely used in the biomedical field, such as constructing microchannels for microfluidic chips.

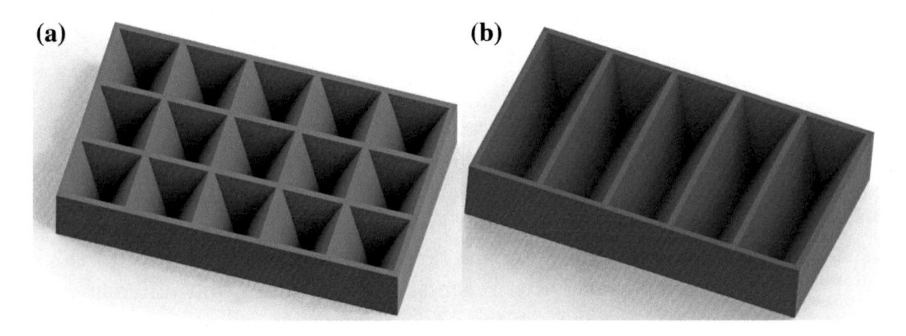

Fig. 2.8 Two types of Si-based microstructures for fabricating micro-/nanohierarchical structures. **a** Inverted pyramid arrays. **b** V-shaped-groove arrays

2.2.2 Fabrication Process

The fabrication process of micro-/nanohierarchical structures (i.e., micro-/nanodual-scale structures) basically contains two steps, i.e., microstructure fabrication and nanostructure fabrication. First, the traditional microfabrication technique is employed to realize microstructures, including photolithography and wet etching. Second, the improved DRIE process proposed above is used to form nanostructures atop microstructures. The fabrication process flow of Si-based micro-/nanohierarchical structures based on the improved DRIE process is shown in Fig. 2.9.

An N-type (100) 4-in. silicon substrate was used, whose thickness is 525 μm and the resistivity is 2–4 Ω cm. First, the silicon substrate was immersed in a 30 % HF solution for one minute to remove the oxidation layer. Then, a ~2000 Å Si_3N_4 layer was deposited on the silicon substrate by low-pressure chemical vapor deposition (LPCVD) process, as shown in Fig. 2.9a. Then, the Si_3N_4 layer was patterned by photolithography and reactive-ion etching (RIE) process, as shown in Fig. 2.9b. To form well-designed microstructures (i.e., inverted pyramid arrays and V-shaped-groove arrays with cross-sectional views of inverted triangle or trapezoid), KOH wet etching process was performed at 85 °C as shown in Fig. 2.9c, followed by removal of Si_3N_4 layer by RIE as shown in Fig. 2.9d. Finally, high-dense high-aspect-ratio nanostructures were formed atop microstructures by an inductively coupled plasma (ICP) etcher using the improved DRIE process. To better describe the advantages of this novel Si-based micro-/nanointegrated

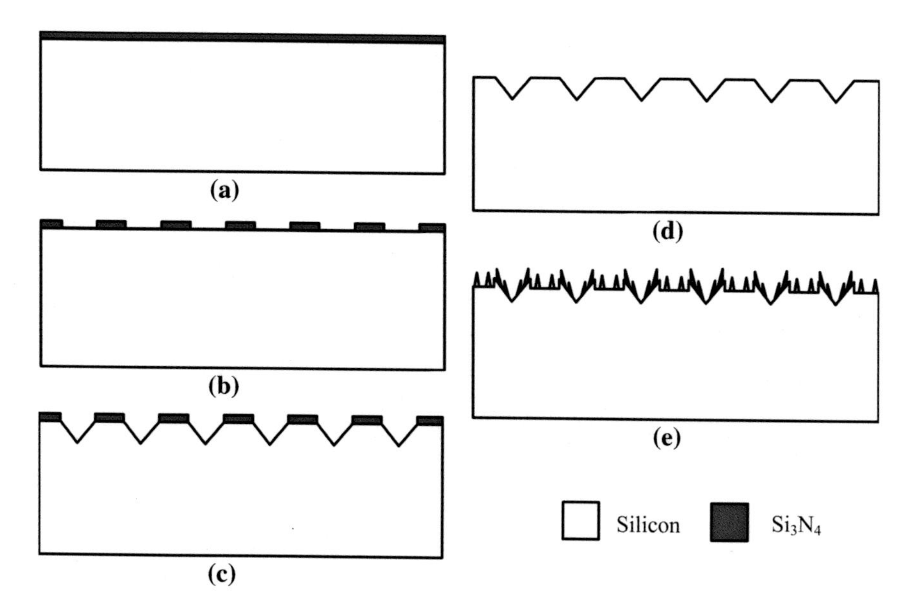

Fig. 2.9 Schematic view of the fabrication process flow of Si-based micro-/nanohierarchical structures based on the improved DRIE process

fabrication technique, we decompose the process in Fig. 2.9e and the detailed illustration of producing nanostructures atop microstructures based on the improved DRIE process is shown in Fig. 2.10.

Figures 2.11 and 2.12 show the SEM images of two types of fabricated Si-based microstructures and micro-/nanohierarchical structures. Figures 2.11I and 2.12I show the morphology of microstructures, i.e., the fabrication result shown in

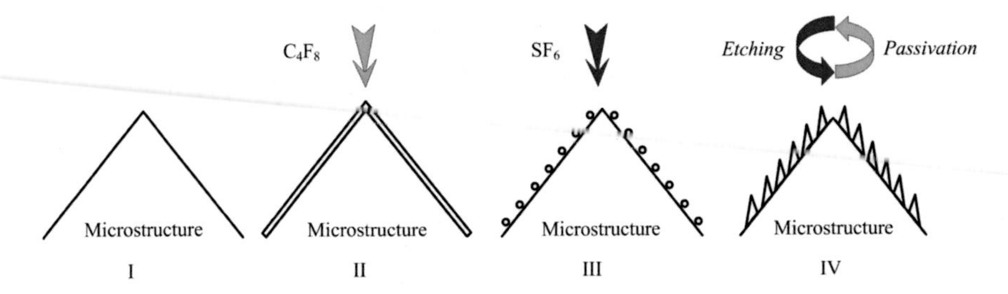

Fig. 2.10 Schematic illustration of producing nanostructures atop microstructures based on optimized DRIE process

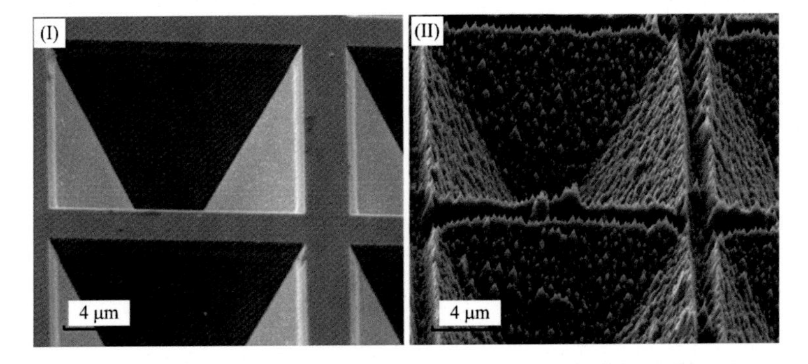

Fig. 2.11 SEM images of (**I**) microstructures and (**II**) micro-/nanohierarchical structures with inverted pyramid arrays

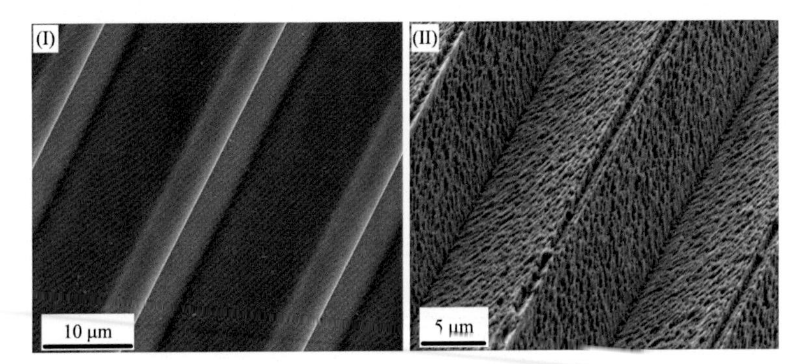

Fig. 2.12 SEM images of (**I**) microstructures and (**II**) micro-/nanohierarchical structures with V-shaped-groove arrays

the Figs. 2.9d, 2.11II and 2.12II shows the micro-/nanohierarchical structures, i.e., the fabrication result as shown in Fig. 2.9e.

2.2.3 Characterization and Analysis of Micro-/NanoHierarchical Structures

In this subsection, all the characterizations of fabricated samples were performed by using scanning electron microscope (Quanta 600F, EFI Company). Figure 2.13 shows the SEM images of fabricated Si-based micro-/nanohierarchical structures. Clearly, according to the top-view image, periodic microstructures are completely covered by nanostructures, even on the inclined surfaces with a title angle of 54.74° shown in Figs. 2.13 and 2.14.

Figure 2.15 shows the SEM images of large-scale micro-/nanohierarchical structures. The fabricated MNHS samples clearly show the perfect uniformity, even at wafer-scale level, which demonstrates the reliability of the presented Si-based micro-/nanointegrated fabrication technique. Regardless of doping type, resistance, and thickness of Si substrates, the fabrication results keep highly consistent. Besides, the periodic microstructures with different dimensions were designed (shown in Table 2.2), and all the SEM images show that microstructures are well covered by nanostructures completely.

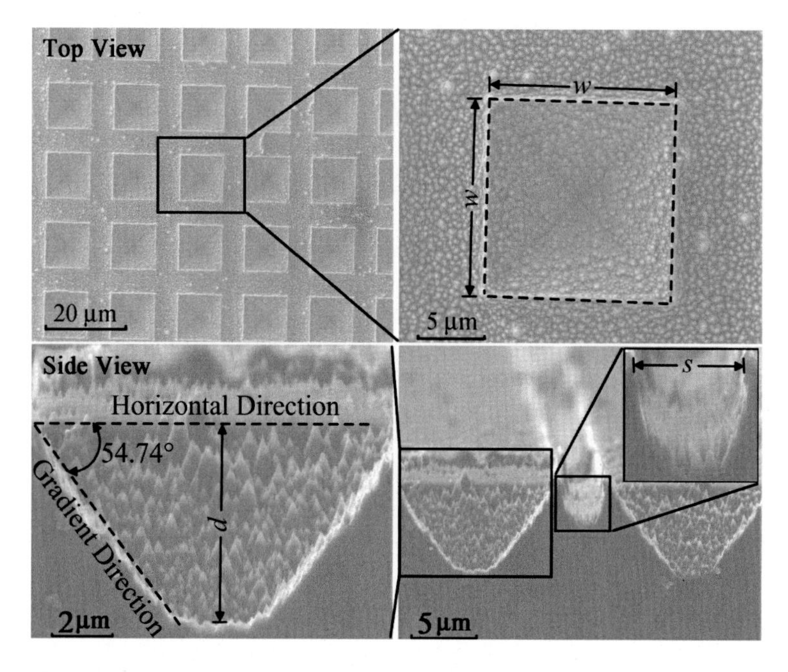

Fig. 2.13 SEM images of Si-based micro-/nanohierarchical structures from top view (*upper*) and cross-sectional view (*lower*)

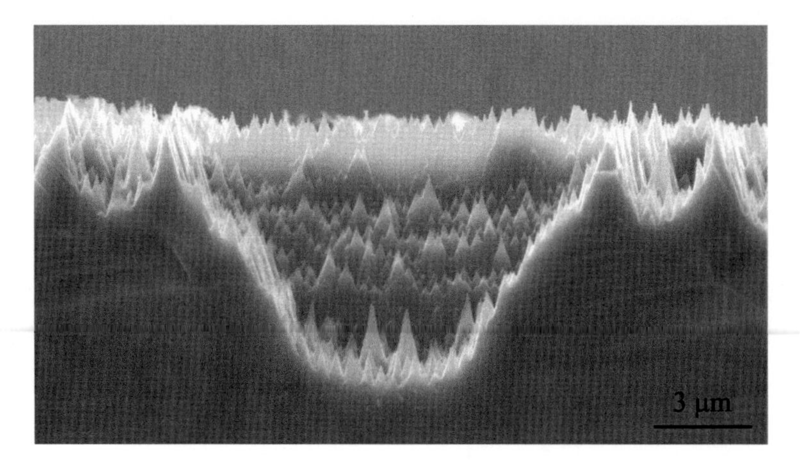

Fig. 2.14 SEM images of Si-based micro-/nanohierarchical structures from cross-sectional view

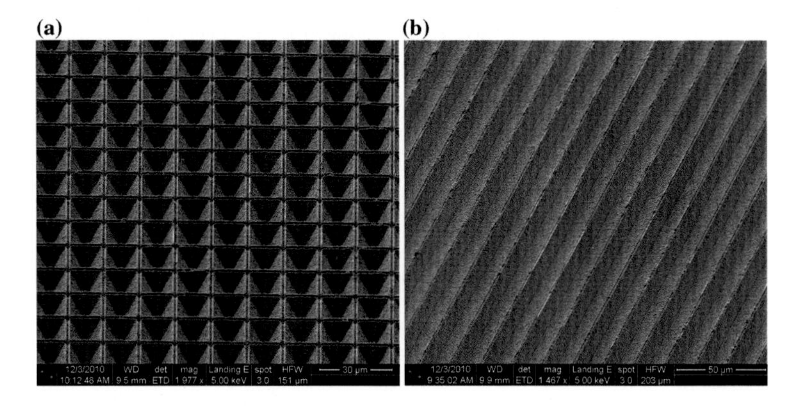

Fig. 2.15 SEM images of large-scale Si-based micro-/nanohierarchical structures. **a** Pyramid-shaped MNHS **b** Groove-shaped MNHS

Table 2.2 Structural parameters of micro-/nanohierarchical structures

Type and No.		Structural parameters		
		Width (μm)	Space (μm)	Depth (μm)
Inverted pyramids	$P_1 \sim P_7$	10, 13, 15, 19	2, 4, 8, 16	4.4
V-shaped grooves	$G_1 \sim G_5$	10, 13, 15	2, 4, 6	13.4

We can realize the microstructures with inverted trapezoids by controlling the KOH etching time shown in Fig. 2.9c and can further realize the micro-/nanohierarchical structures with the cross-sectional view of inverted trapezoid shown in Fig. 2.16. This also validates the universality of this fabrication technique. As mentioned above, regardless of doping type, resistance, and thickness of

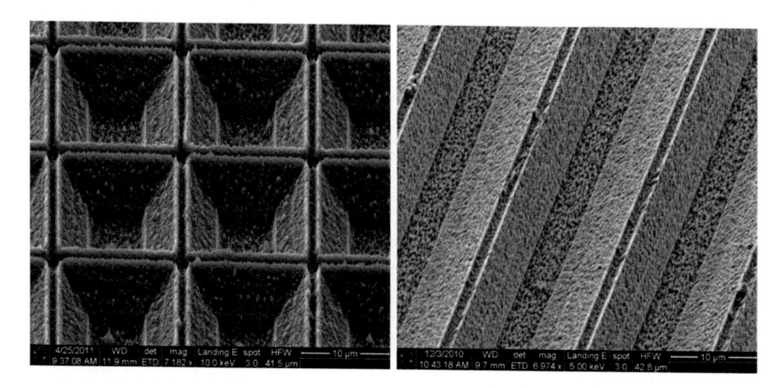

Fig. 2.16 SEM images of Si-based micro-/nanohierarchical structures with the cross-sectional view of inverted trapezoid

Si substrates and regardless of profiles and sizes of microstructures on the surface, this Si-based micro-/nanointegrated fabrication technique based on the improved DRIE process is repeatable, controllable, cost-efficient, and reliable.

2.3 Interaction of Multiscale Structures Based on Silicon

Due to the combination of multiscale structures, more factors have to be considered in the fabrication procedure, and the related chemical–physical energy transfer becomes more complicated, i.e., the combination of multiscale structures brings some new challenges and phenomena, which are an essential research field of micro-/nanointegrated fabrication. However, there is a lack of knowledge considering multiscale interaction because of size effect, controllability, and repeatability.

In order to investigate the interaction of structures at different scales, the microstructures are specifically designed to be inverted pyramids and V-shaped grooves, and each type contains two cross sections with inverted triangle and trapezoid, respectively, shown in Figs. 2.11, 2.12, 2.15 and 2.16. For different micro-/nanohierarchical structures, their basic unit can be classified into two types, horizontal and inclined surfaces, as shown in Fig. 2.17, where the solid line and the dash line show the profile of microstructures and MNHS, respectively.

During the MNHS fabrication, the etching rate on the horizontal surface is faster than that on the inclined surface, as shown in Fig. 2.18. In the fixed process time of t, the depth change of horizontal surfaces is set as ΔH and that of inclined surfaces is set as Δh, thus $\Delta H > \Delta h$ according to Figs. 2.14, 2.17 and 2.18. And there is a quantitative relation as follows:

$$\Delta H(= \Delta h + \Delta d) > \Delta h \quad (2.3)$$

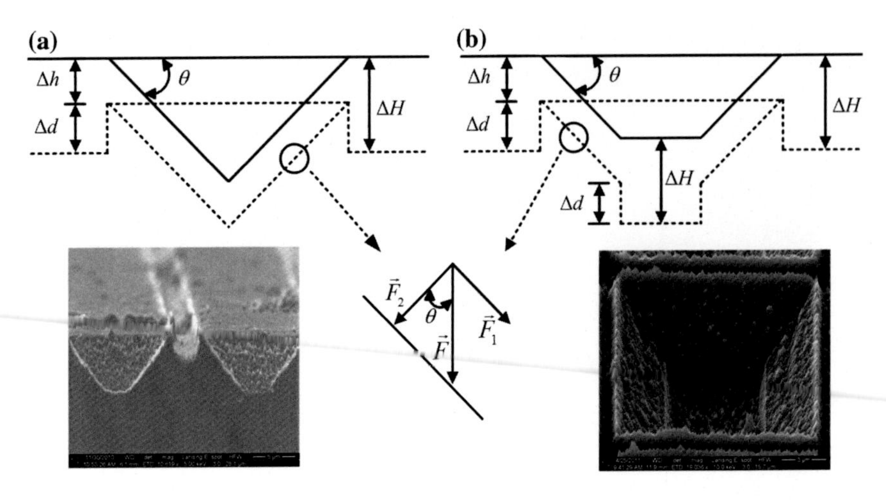

Fig. 2.17 Illustration of the formation of micro-/nanohierarchical structures (*solid line* the profile of microstructure; *dash line* the profile of MNHS). **a** Triangle cross section. **b** Trapezoid cross section

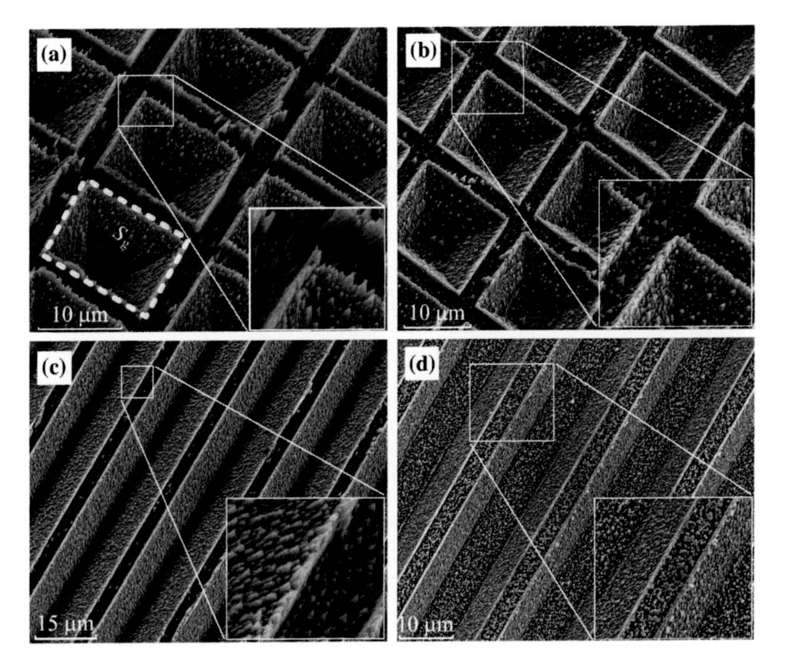

Fig. 2.18 SEM images of Si-based micro-/nanohierarchical structures covered by a thin gold layer

By defining the corresponding average etching velocities $V_h = \Delta H/t$ and $V_i = \Delta h/t$, we can conclude the following relation with the same etching time.

$$V_h > V_i \tag{2.4}$$

The above difference can be attributed to energy flux vector field of plasma ions. There are two RF sources with one controlling the direction of electric field. When the platen power is fixed, the energy flux vector field of ion beam F in the vertical direction will keep constant. The F, whose direction is normal to the horizontal surfaces, can be decomposed into two orthogonal components, i.e., F_1 and F_2, which are parallel and normal to the inclined surface, respectively, as shown in Fig. 2.18.

$$F_1 = \mathrm{F} \times \sin\theta \tag{2.5}$$

$$F_2 = \mathrm{F} \times \cos\theta \tag{2.6}$$

And therefore, the etching velocity on the inclined surface is smaller than that on the horizontal surface due to $F_2 < F$. Moreover, the inclined angle θ is approximately constant with the value of 54.74° during the formation of MNHS. Therefore, the relation between V_h and V_i exists as follows:

$$V_h = f(V_i, \theta) \tag{2.7}$$

We have investigated the dual-scale coupling effect during the formation of MNHS above, and the morphology of MNHS also significantly affects the further fabrication. To explore this coupling effect in the multiscale structures, we fabricated Si-based MNHS covered by a thin gold layer of ~500 Å by sputtering process, as shown in Fig. 2.18. The surface of Si-based micro-/nanohierarchical structures in Fig. 2.18 presents regular bright and dark areas with the bright on the inclined surface and the dark on the horizontal surface and at the bottom of grooves. To better illustrate these phenomena, Fig. 2.19 gives SEM images of part-enlarged view.

According to the SEM images of micro-/nanomultiscale structures shown in Fig. 2.19, gold particles can only be sputtered on inclined surfaces (i.e., C zone). In other words, there are no gold particles on horizontal surfaces (i.e., A zone) and at the bottom of microstructures (i.e., B zone). We believe that two reasons result in these phenomena. First, the dissymmetrical nanostructures atop inclined surfaces are more favorable to capture sputtered gold particles due to the sharper sidewall, which is almost parallel to the incident particles and can reduce the probability of particle rebound. Second, the surface substances on the nanostructures in the C zone are different from those in the A/B zones due to the difference in the energy flux vector field of ion beams and the electric field influence of small-size microstructures during the formation of nanostructures onto microstructures. As for the electric field influence of small-size microstructures, these phenomena become less obvious when the size of the microstructures gradually increase, as shown in Fig. 2.18d.

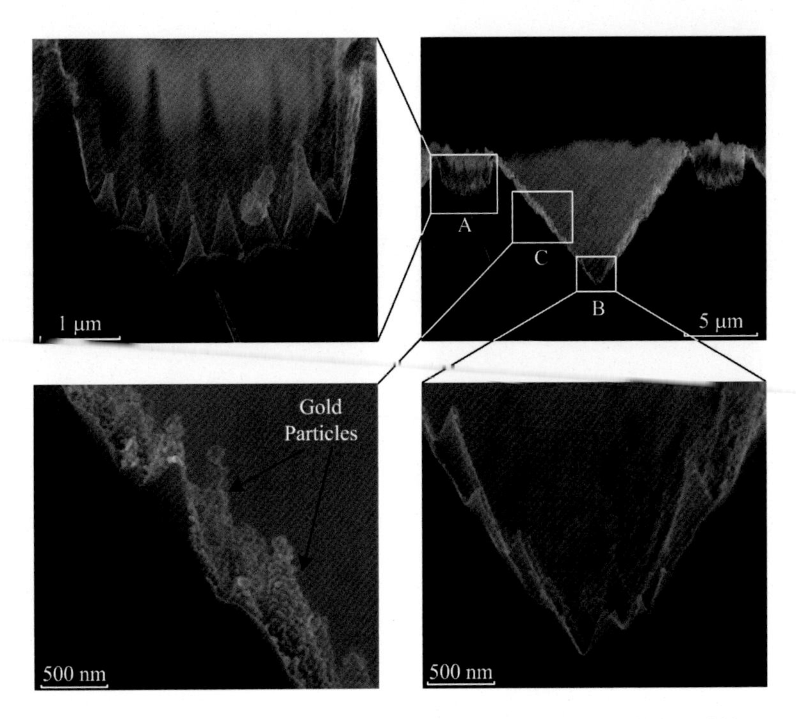

Fig. 2.19 SEM images of cross-sectional view of Si-based MNHS covered by a gold layer

2.4 Properties of Si-Based Micro-/NanoHierarchical Structures

Si-based materials have widespread use for optical devices and photoelectric converters; however, the high reflectance hampers their further development. Increasing the roughness has been proven as an effective method to reduce the reflectance, and thus, Si-based MNHS combining periodic microstructures and high-dense nanostructures can greatly lower the reflectance, which is very useful for improving the sensitivity and photoelectric conversion efficiency of Si-based optical devices. Besides, such a multiscale combination also significantly maximizes the surface area and minimizes the solid–liquid contact area of a water droplet on the surface, thus improving the super-hydrophobic property. Moreover, due to using C_4F_8 as the passivation gas for DRIE process, the fluorocarbon polymer residues on the Si substrate surface will significantly reduce the surface energy and further strengthen the super-hydrophobicity. Thus, this subsection is to introduce two excellent properties of Si-based MNHS, i.e., wide-band anti-reflectance and super-hydrophobicity.

2.4.1 Anti-reflective Property

The roughness of micro-/nanohierarchical structures is greatly improved due to the combination of microstructures and nanostructures. Therefore, when light is incident on the dual-scale surface, the incident light is repeatedly reflected and refracted, thus realizing the anti-reflectance. Figure 2.20 shows the reflectance measurement results among samples including micro-/nanohierarchical structures (MNHS), black silicon (BS), and microstructures (MS) under an incidence angle of 6° using Lambda 950 spectrophotometer (Perkin Elmer Inc.). Clearly, with the same DRIE process parameters, the reflectance order is as follows:

$$R_{\mathrm{MNHS}} < R_{\mathrm{BS}} < R_{\mathrm{MS}} \tag{2.8}$$

Compared with flat silicon substrate, pure microstructures have lower optical reflectance but still high ($R > 30\,\%$). The reflectance of pure microstructures increases as the wavelength of incident light decreases and is higher than 5 % in the visible range. Pure nanostructures have good wideband anti-reflectance, and the reflectance decreases to 1 % in the measurement range. And MNHSs further reduce the reflectance. The reflectance of MNHSs without any reflective coating has been reduced to less than 0.6 % in a wavelength range from 200 to 2500 nm, and correspondingly, the optimized size parameters, i.e., width = 13 μm and space = 4 μm, are obtained. Besides, MNHS with inverted pyramids exhibit lower reflectance than with V-shaped grooves, due to the more dense array structures of the former.

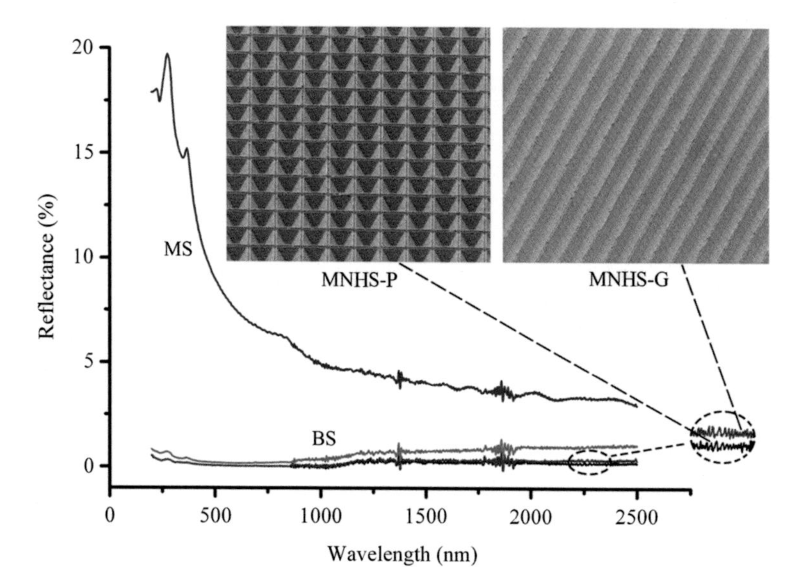

Fig. 2.20 Comparison of reflectance among MNHSs with inverted pyramids and V-shaped grooves under an incidence angle of 6°. *BS* black silicon; *MS* microstructures; *MNHS-P* MNHS with inverted pyramids; *MNHS-G* MNHS with V-shaped grooves

For the rough surfaces, pure reflectance test does not accurately reflect the true anti-reflective effect due to the presence of the scattering phenomena. Therefore, we further measured hemispherical reflectance of fabricated samples, i.e., the sum of reflectance (*R*) and scattering (*S*), shown by dash line in Fig. 2.21. The hemispherical reflectance results indicate the same trend as pure reflectance results. The MNHS have the best anti-reflectance property with hemispherical reflectance lower than 5 %. Compared with the polished normal silicon surface, whose hemispherical reflectance is larger than 35 %, the MNHSs have suppressed the hemispherical reflectance by a factor of >85.7 %, showing a highly effective optical absorption characteristics.

To comprehensively test the optical properties of MNHS samples, we also investigated their optical absorption (*A*). The optical absorption can be calculated according to the following equation,

$$A = 100\,\% - R - S - T \tag{2.9}$$

where *A* denotes absorption, *R* denotes reflectance, *S* denotes scattering, and *T* denotes transmittance.

As mentioned above, the hemispherical reflectance is the sum of *R* and *S*. That is to say, after measuring the hemispherical reflectance and the transmittance, the optical absorption can be directly obtained, as indicated by the solid line in Fig. 2.21. The MNHSs can absorb ~95 % incident light during the whole solar spectrum (i.e., 400–1000 nm), making them an excellent optical materials.

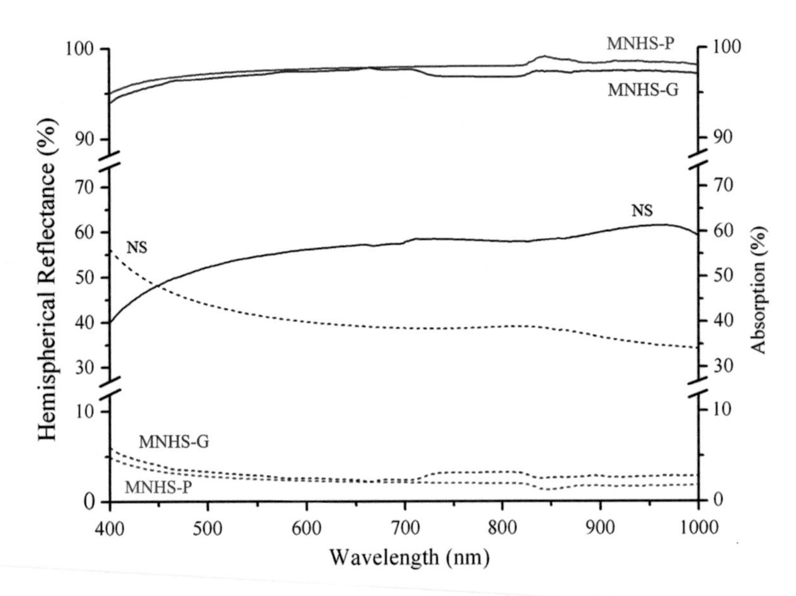

Fig. 2.21 Hemispherical reflectance spectrums (*dash lines*) and absorption spectrums (*solid lines*) of optimized MNHS. *NS* normal silicon; *MNHS* micro-/nanodual-scale structure (*MNHS-P* MNHS with inverted pyramids; *MNHS-G* MNHS with V-shaped-groove arrays)

For pure microstructures without nanostructures, their optical properties are sensitive to their structural parameters. Therefore, the structural optimization and accurate fabrication are the key for their applications in the optical field, but in the meantime, this also leads to their reliability descending and cost increasing. The inclined surfaces of microstructures can effectively reflect the incident light repeatedly, thus resulting in the high optical absorption. While most of the incident light on the horizontal surface will be directly reflected out and wasted. And therefore, the ratio of surface area in the horizontal direction to the total surface area on a sample, which is defined as *RSA*, qualitatively determines the quality of its optical property. That is to say, the larger the *RSA*, the higher the hemispherical reflectance; the smaller the *RSA*, the lower the hemispherical reflectance. For pure microstructures, in order to achieve the most excellent anti-reflectance property, the *RSA* value needs to be further reduced. However, the minimum value of *RSA* is restricted by the minimum width of photolithography. Let us take Figs. 2.11 and 2.12 as examples, the space between two adjacent microstructures is determined and limited by the minimal line width of photolithography. The micro-/nanohierarchical structures can be employed to overcome this drawback due to the combination of wideband anti-reflective nanostructures with underlying microstructures.

As shown in Figs. 2.13 and 2.18, to facilitate illustration and understanding, we define the area of a single inverted pyramid or groove as S_g and assume that there are n-inverted pyramids/grooves on the whole sample. Then, we can calculate *RSA* based on the following equation:

$$\mathrm{RSA} = (S_{\mathrm{total}} - n \times S_g)/S_{\mathrm{total}} \tag{2.10}$$

where S_{total} is 2 cm × 2 cm. The hemispherical reflectance of MNHS was measured by a UV-3600 spectrophotometer (SHIMADZU Corporation), and the results are shown in Fig. 2.22. Clearly, the hemispherical reflectance of pure microstructures increases from 13.2 to 24 % as the *RSA* increases from 41.6 to 61.7 %. In contrast, the hemispherical reflectance variation of MNHS is limited to less than 2 %, although the *RSA* changes during a wide range from 13.4 to 79.9 %. These results show that the hemispherical reflectance of MNHS is very stable even though the structural parameters vary a lot. Therefore, the optical reflectance of MNHS is not sensitive to structural parameters, which makes it more tolerant toward process errors and solves the mentioned limitation of photolithography.

2.4.2 Super-hydrophobic Property

Hydrophobic materials have attracted much attention in the past decades due to the widespread applications in the industry production and the fundamental research [5, 6]. The hydrophobicity of low-surface-energy materials can be enhanced by surface roughening [7, 8]. Nanostructures, such as nanoporous, nanowires, and nanopillars, have been used to roughen the surfaces and side walls

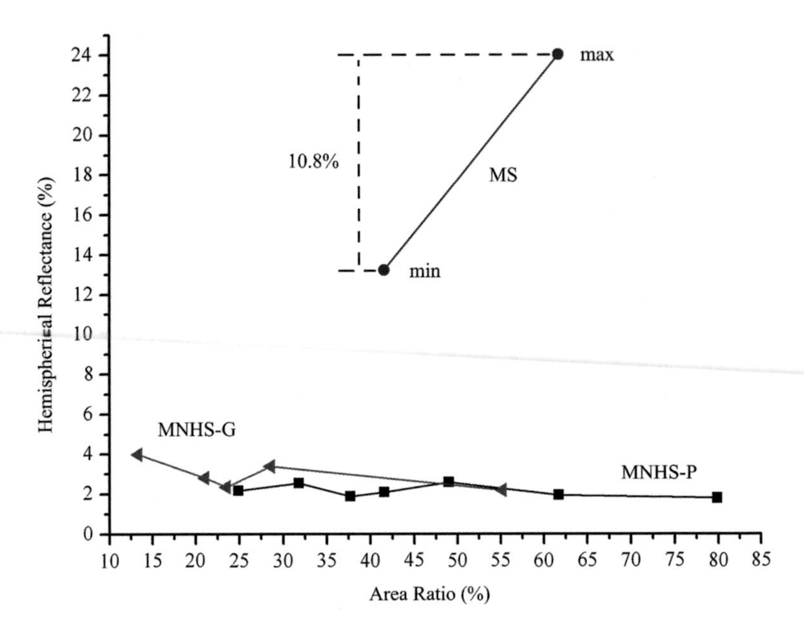

Fig. 2.22 Relation between the hemispherical reflectance and the surface area ratio (RSA) of horizontal surface to the total surface under an incidence wavelength of 700 nm. *MS* pure microstructure; *MNHS-P* MNHS with inverted pyramids; *MNHS-G* MNHS with V-shaped grooves

of hydrophobic materials to realize the super-hydrophobicity [9, 10]. Although several fabrication methods have been proposed to realize the nanostructures on the microstructures, they are limited by some problems more or less, such as requiring mask, containing several steps, disability in fabricating nanostructures on side walls and low efficiency, as mentioned in Sect. 2.1.2.

While the improved DRIE process for micro-/nanohierarchical fabrication shows its unique properties, such as maskless, wafer level, and mass production, to respond the above problems, the fluorocarbon polymer deposited by passivation steps significantly reduces the surface energy of silicon substrate and thus further strengthens the wettability and realizing the super-hydrophobicity with contact angle (CA) larger than 150° and rolling angle (RA) less than 1°. Here, the static CA was measured by using OCA20 video-based contact angle meter (DataPhysics Instruments GmbH), and other liquid dynamics properties were measured by using JC2000D contact angle meter (POWEREACH).

Figure 2.23 shows the contact angle measurement results of Si-based micro-/nanohierarchical structures. Pyramid-shaped MNHS (MNHS-P) samples with cross sections of inverted triangle and trapezoid are shown in Fig. 2.23a, b, respectively, and groove-shaped MNHS (MNHS-G) samples with cross sections of inverted triangle and trapezoid are shown in Fig. 2.23c, d, respectively. Generally, MNHS-P samples achieve higher CAs than those of MNHS-G samples. Meanwhile, the sample with inverted-triangle cross section exhibits higher CA than that with trapezoid cross section. With optimal structural parameters

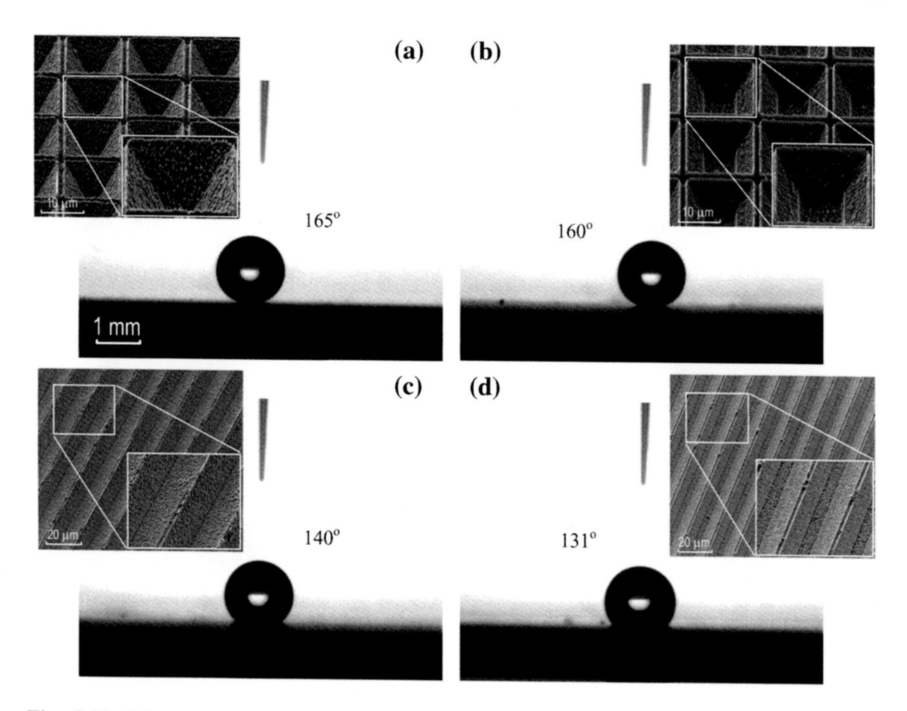

Fig. 2.23 Measurement results of static contact angles of Si-based micro-/nanohierarchical structures (water droplet volume = 2 μL): **a** ~165°; **b** ~160°; **c** ~140°; **d** ~131°

(i.e., $w = 13$ μm, $s = 2$ μm, and $d = 9.2$ μm), the MNHS-P with inverted-triangle cross section achieved the largest CA of 165°, as shown in Fig. 2.23a.

We further measured the rolling angle, which also characterizes the sample's hydrophobicity. The sample was placed on a flat platform with an inclined angle of below 1°, and a 10 μL water droplet was dropped on the sample surface and then freely rolled along the surface, shown in Fig. 2.24. Therefore, the RA of optimized MNHS is below 1°. According to published literatures, the Cassie's model can be used to describe the wettability of the material with ultra-large CA and extreme-small RA. Furthermore, ultra-large CA and extreme-small RA make these MNHS very useful for super-hydrophobic applications, especially to realize the self-cleaning surface. When the rain drops on these MNHS surfaces, it will flow away and take dusts and particles away.

The above CA and RA measurement results confirmed the excellent super-hydrophobic property of Si-based MNHS. As is known, the reliability and stability of super-hydrophobic materials are very important for the practical applications. We further performed squeezing and impact tests to verify the reliability and stability of the super-hydrophobicity of MNHS, as shown in Figs. 2.25 and 2.26. A 10 μL water droplet suspended on the tip of an injector, while the fabricated sample touched and squeezed the droplet. Sequentially, the sample descended and departed from the droplet, but the water droplet still suspended on the tip, shown

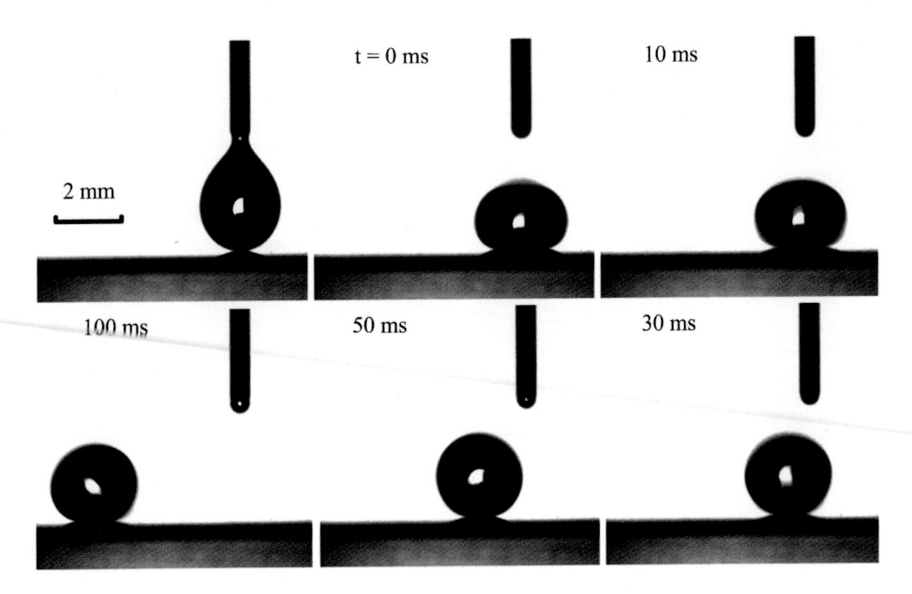

Fig. 2.24 Sequential images of a water droplet (volume = 10 μL) sliding on the MNHS sample with a tilted angle of below 1°

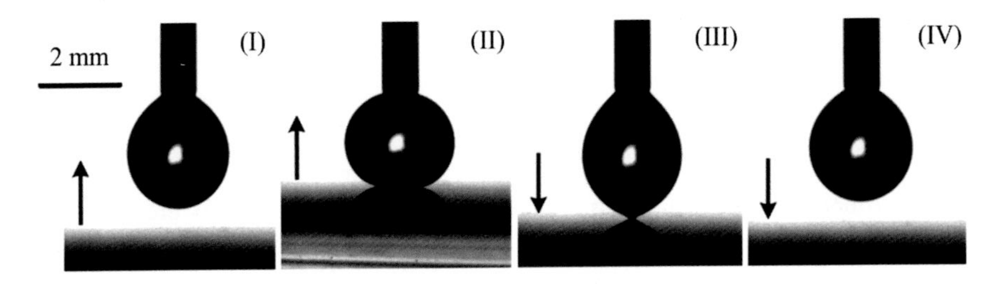

Fig. 2.25 Sequential images of squeezing test (water droplet volume = 10 μL)

in Fig. 2.25. This shows the interface force between the sample and the water droplet is very weak, which validates the stability of the super-hydrophobicity of MNHS.

In Fig. 2.26, a 15 μL water droplet dropped from a height of 5 mm with initial speed of 0 m/s and then impacted the surface of MNHS. According to the sequential images at the 20, 40, 70, 120, and 160 ms, the water droplet bounced for 4–5 times until the potential energy was completely consumed by the friction and water internal flow. In the whole impact procedure, the water droplet did not adhere to the sample surface, demonstrating the reliability of super-hydrophobicity of MNHS. It is worth mentioning that the above reliability and stability tests were carried out three months after the samples were fabricated, which further proves the long-term stability of the super-hydrophobicity of MNHS.

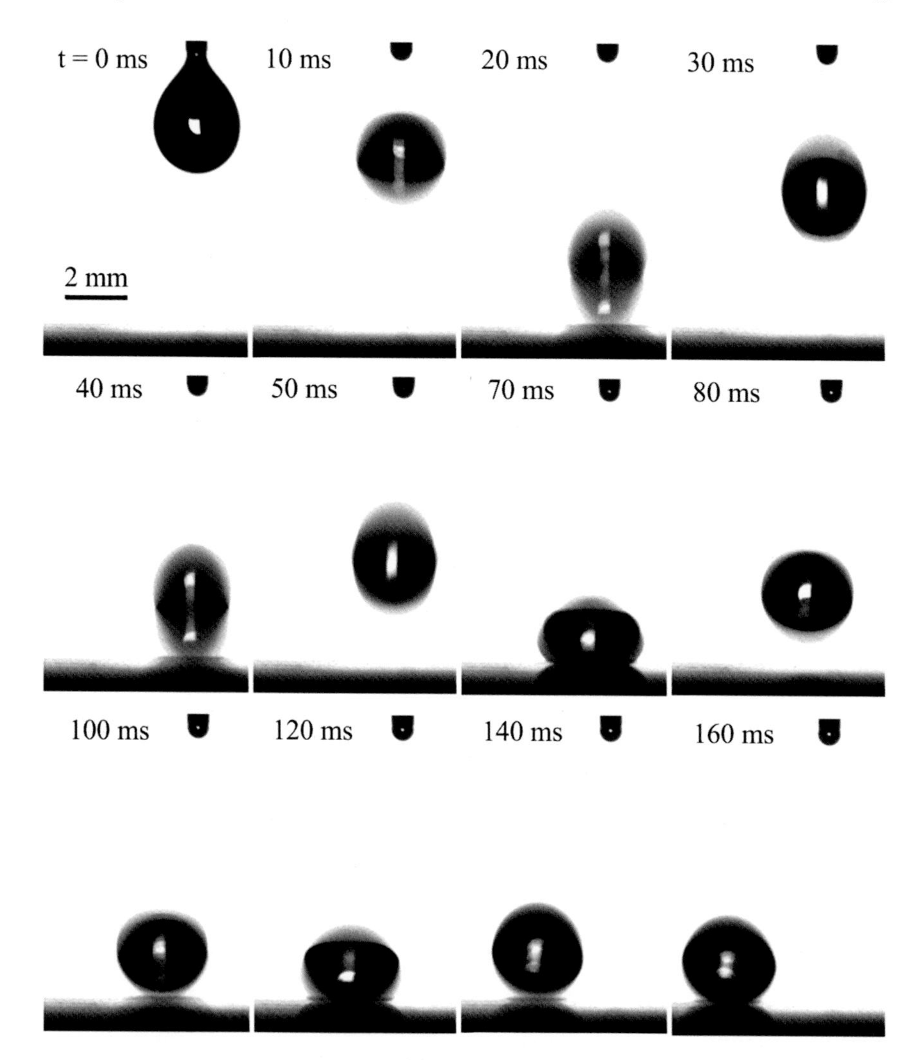

Fig. 2.26 Sequential images of a water droplet (volume = 15 μL) impacting the MNHS sample

Electrowetting is an important and useful property for hydrophobic materials in many applications, such as display devices [11–13]. However, in other fields, e.g., electricity-driven micro-/nanofluids, the reliable super-hydrophobicity in the high voltage is essential. Combining a Picoammeter/Voltage Source (6487, Keithley Instruments Inc.) with the contact angle meter, we investigated the electrowetting property of MNHS by plotting the curve of contact angle versus applied voltage with a 10 μL water droplet, shown in Fig. 2.27. Obviously, the contact angle of MNHS is kept constant at 158° during the voltage increasing from 0 to 300 V, while it decreases to 143° and 115° under the voltage of 350 and 400 V, respectively. And therefore, the electrowetting threshold voltage of the fabricated MNHS

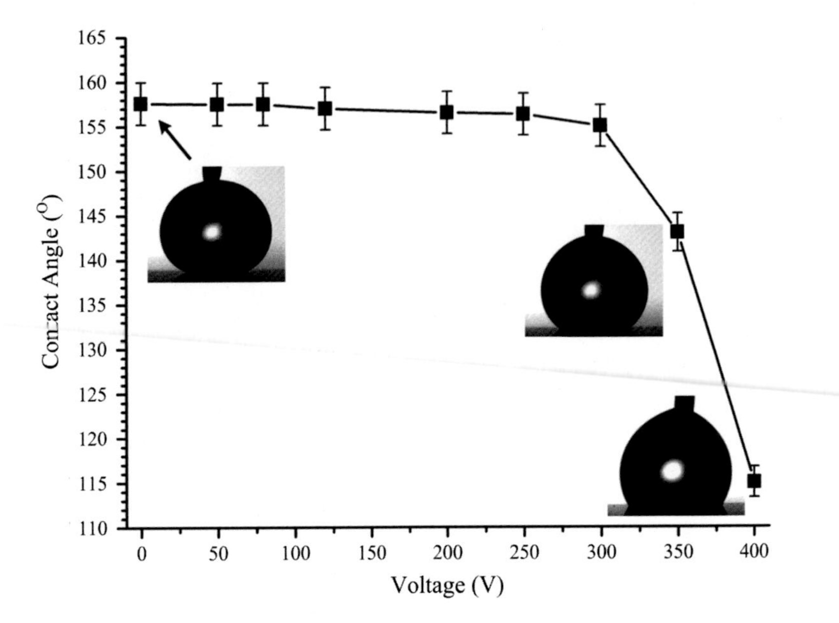

Fig. 2.27 Contact angle of water droplet on the surface of micro-/nanohierarchical structures versus applied voltage (water droplet volume = 10 μL)

is ~300 V, higher than that of normal electrowetting materials (below 200 V). In one word, the fabricated MNHS samples based on the optimized DRIE process showed a stable super-hydrophobic property.

2.5 Conclusions

In this chapter, by optimizing process parameters, we propose a maskless nanofabrication method based on an improved DRIE process. Four working ranges of DRIE process were figured out by lots of independent and comparative experiments. The subsequent experiments are carried out to study the key process parameters of DRIE process and their qualitative and quantitative effects on the profile of nanostructures. Finally, a maskless nanofabrication process for silicon is successfully developed, which shows advantages of controllability, large-scale, reliability, and repeatability. By combining traditional microfabrication techniques, we propose a novel micro-/nanointegrated fabrication technique, and several types of Si-based micro-/nanohierarchical structures are successfully realized. The interaction effects of multiscale structures during the micro-/nanointegrated fabrication procedure is observed, including the interaction between microstructures and nanostructures and the influence of the MNHS geometry on the gold deposition. According to the optical property and liquid property measurements, the fabricated Si-based MNHS materials show excellent anti-reflectance and super-hydrophobicity.

In summary, by experimental exploration, theoretical analysis, and mechanism investigation, we propose and realize a Si-based micro-/nanointegrated fabrication technique based on an improved DRIE process, which is cost-efficient, mass-fabrication, and wafer-level.

References

1. X.S. Zhang, Q.L. Di, F.Y. Zhu, G.Y. Sun, H.X. Zhang, Wideband anti-reflective micro/nano dual-scale structures: fabrication and optical properties. Micro Nano Lett. **6**, 947–950 (2011)
2. G. Sun, T. Gao, X. Zhao, H. Zhang, Fabrication of micro/nano dual-scale structures by improved deep reactive ion etching. J. Micromech. Microeng. **20**, 075028 (2010)
3. X.S. Zhang, F.Y. Zhu, G.Y. Sun, H.X. Zhang, Fabrication and characterization of squama-shape micro/nano multi-scale silicon material. Sci. China E **55**, 3395–3400 (2012)
4. X.S. Zhang, Q.L. Di, F.Y. Zhu, G.Y. Sun, H.X. Zhang, Superhydrophobic micro/nano dual-scale structures. J. Nanosci. Nanotechnol. **13**, 1539–1542 (2013)
5. X. Zhang, F. Shi, J. Niu, Y. Jiang, Z. Wang, Superhydrophobic surfaces: from structural control to functional application. J. Mater. Chem. **18**, 621–633 (2008)
6. N. Verplanck, Y. Coffinier, V. Thomy, R. Boukherroub, Wettability switching techniques on superhydrophobic surfaces. Nanoscale Res. Lett. **2**, 577–596 (2007)
7. L. Feng, Y. Song, J. Zhai, B. Liu, J. Xu, L. Jiang, D. Zhu, Creation of a superhydrophobic surface from an amphiphilic polymer. Angew. Chem. Int. Ed. **115**, 824–826 (2003)
8. A. Lafuma, D. Quéré, Superhydrophobic states. Nat. Mater. **2**, 457–460 (2003)
9. C. Lee, C.J. Kim, Maximizing the giant liquid slip on superhydrophobic microstructures by nanostructuring their sidewalls. Langmuir **25**, 12812–12818 (2009)
10. B.S. Kim, S. Shin, S.J. Shin, K.M. Kim, H.H. Cho, Micro-nano hybrid structures with manipulated wettability using a two-step silicon etching on a large area. Nanoscale Res. Lett. **6**, 333 (2011)
11. H. You, A.J. Steckl, Three-color electrowetting display device for electronic paper. Appl. Phys. Lett. **97**, 023514 (2010)
12. F. Mugele, J.C. Baret, Electrowetting: from basics to applications. J. Phys. Condens. Matter **17**, R705–R774 (2005)
13. W.C. Nelson, C.J. Kim, Monolithic fabrication of EWOD chips for picoliter droplets. J. Microelectromech. Syst. **20**, 1419–1427 (2011)

Chapter 3
Micro-/Nanointegrated Fabrication Technique for Flexible Materials

Abstract In Chap. 2, we introduce a mass-production Si-based micro-/nanointegrated fabrication technique, which is simple, cost-efficient, and compatible with CMOS process. The fabricated samples show two important properties of wide-band anti-reflectance and stable super-hydrophobicity; thus, this novel Si-based micro-/nanointegrated fabrication technique has the attractive application potential. However, silicon is fragile and bio-incompatible, which hinds the further application of this technique in biomedical field. Therefore, we extend this micro-/nanointegrated fabrication technique from silicon to flexible materials, especially for biocompatible flexible materials, and then, a universal micro-/nanointegrated fabrication technology suitable for many materials is developed.

As is mentioned previously, due to the combination of microstructures and nanostructures, micro-/nanohierarchical structures possess many attractive properties. Additionally, flexible materials have the outstanding flexibility and biocompatibility, which are widely used in microfluidics and biomedical fields. Thus, micro-/nanohierarchical structures based on flexible materials have a bright potential future. However, the traditional fabrication processes contain at least two steps (i.e., microfabrication and nanofabrication), which are complicated. In addition, the fabrication precision is not ideal at nanoscale level. Therefore, in order to overcome these drawbacks, we proposed a single-step fabrication technique based on the combination of ultra-low-surface-energy silicon mold and replication process of flexible materials to realize micro-/nanohierarchical structures with flexible materials. This single-step technique is simple, low cost, and of high-throughput, and more importantly, the fabricated samples show several remarkable properties.

X.-S. Zhang, *Micro/Nano Integrated Fabrication Technology and Its Applications in Microenergy Harvesting*, Springer Theses, DOI 10.1007/978-3-662-48816-4_3

3.1 Replication Process and Surface-Energy Control

3.1.1 Replication Process Based on Silicon Mold

Flexible materials are widely used in many fields resulting from their excellent flexibility and stretchability. Unfortunately, this excellent flexibility also brings the challenge to controllably fabricate high-precision patterns on the surfaces of flexible materials. In order to respond to this limitation, researchers developed a simple fabrication approach named replication process (i.e., casting mold process). Replication process is based on self-leveling effect and conformal effect, which can be briefly summarized into three steps, including covering prepatterned mold with liquid flexible materials, subsequently solidifying liquid flexible materials, and eventually peeling off solid flexible materials from the mold. Therefore, by the combination of the Si-based micro-/nanointegrated fabrication technique present in Chap. 2 with the replication process of flexible materials, we can replicate micro-/nanohierarchical structures from prepatterned silicon mold to flexible materials.

As is shown in Fig. 3.1, micro-/nanohierarchical structures on flexible materials are complementary to patterns on the silicon mold. In detail, microscale inverted pyramids on silicon mold are transferred to be microscale pyramids on flexible materials, while nanoscale pillars are transferred to be nanoscale porous. The most important key point of replication process is that the flexible material tightly contacts with the silicon mold at the interface and that patterns transferred to the flexible material are not damaged during peeling off. Thus, several techniques are developed to realize the above target, such as pumping for vacuum and applying high pressure and high temperature. But, these techniques also bring another challenge, i.e., bonding between silicon mold and flexible materials. Taking the common flexible material of PDMS as an example, baking process is necessary for the solidification of PDMS. However, when the temperature is in the range of

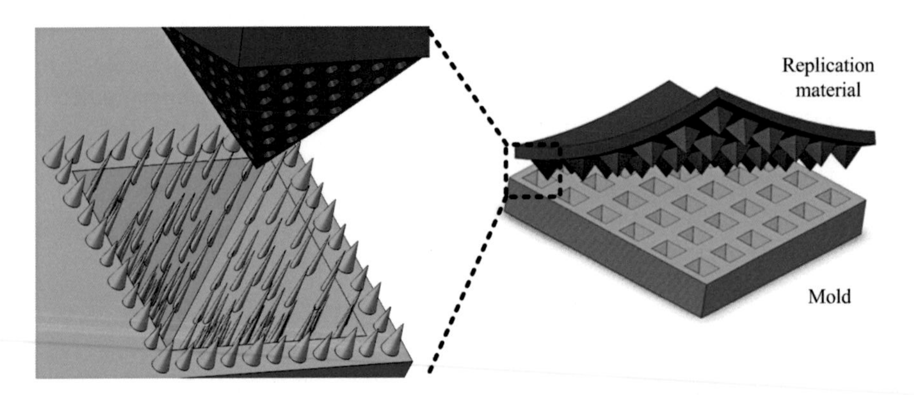

Fig. 3.1 Schematic view of pattern transfer from mold to flexible material

80–90 °C, PDMS film will be bonded with silicon mold and it is very difficult to separate them. In order to solve this problem, the surfactant is used to cover the silicon mold in advance and prevent the interface bonding [1, 2].

In principle, the function of the surfactant reduces the surface energy of mold, and then, the interface force between flexible materials and mold is sharply weakened. Unfortunately, the surfactant restricts the widespread use of replication process due to two main reasons. First, the surfactant residuals may pollute surfaces of the flexible material and the mold. Although organic solutions and the ultrasonic cleaning process are usually used to remove surfactant residuals, actually it is hard to completely remove them and the surface of flexible material may be damaged. Second, the surfactant covers the surface of mold, and thus, parts of patterns are covered, which reduces the precision of pattern transfer. Especially for nanoscale structures, they are fully covered by surfactant and disappear during the replication process. Therefore, the surfactant has already become one of the biggest challenges for traditional replication process of flexible material.

Considering the fundamental mechanism of surfactant, we believe that sharply reducing the surface energy and realizing an ultra-low-surface-energy mold will be very helpful for the replication process of flexible material, which makes the surfactant unnecessary. Thus, the controllable fabrication of ultra-low-surface-energy molds is the key point to break through the limitation of traditional replication technique.

3.1.2 Controllable Fabrication of Ultra-Low-Surface-Energy Silicon Mold

Surface energy is also named interface energy, which quantifies the disruption of intermolecular bonds that occur when a surface is created. If both temperature and pressure are constant, surface energy equals surface free energy. The wettability of surface is tightly related to its surface energy, which can be used to qualitatively estimate the surface energy. Generally, higher surface energy means stronger hydrophilicity, while lower surface energy means stronger hydrophobicity. Thus, from the experimental point of view, realizing ultra-low-surface-energy materials means fabricating super-hydrophobic materials.

According to the analysis in Sect. 2.4.2, the improved DRIE process can be used to fabricate super-hydrophobic Si-based micro-/nanohierarchical structures, and the main reasons can be summarized as follows: First, the combination of microstructures and nanostructures minimized the liquid–solid interface area; second, the fluorocarbon polymer deposited by passivation steps sharply reduces the surface energy. However, this fluorocarbon polymer exists in the form of residuals due to the etching effect of etching steps which is non-uniform and it is hard to cover the silicon substrate completely. As is mentioned in Sect. 2.1, the equipment of ICP for DRIE process can easily control etching and passivation steps.

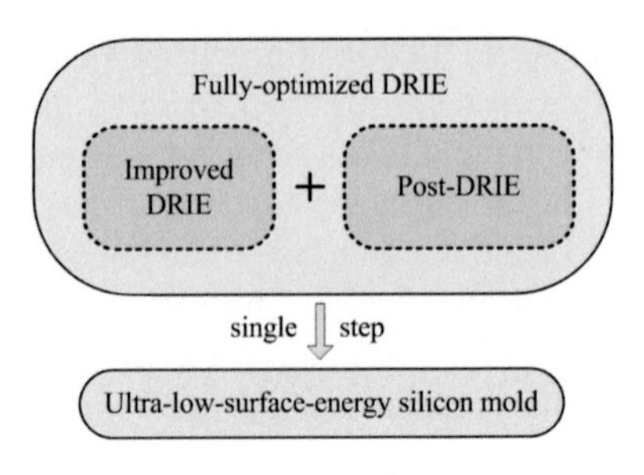

Fig. 3.2 Illustration of the fully optimized DRIE process

Thus, a further improved DRIE process was developed, which consists of the improved DRIE process and the post-DRIE process. Here, the post-DRIE process contains only passivation steps, which can be utilized to form a nanoscale fluorocarbon polymer atop silicon substrate. In order to simplify the explanation, this further improved DRIE process is called fully optimized DRIE process, as shown in Fig. 3.2. Although this fully optimized DRIE process is divided into two parts, actually it is a single-step fabrication process. With one optimized process recipe, we can fabricate Si-based micro-/nanohierarchical structures covered by a uniform fluorocarbon polymer. In detail, we only need to place microstructured silicon samples into the chamber of ICP, and then, ultra-low-surface-energy silicon molds with micro-/nanohierarchical structures are obtained directly by using this fully optimized DRIE process.

As a result of the minimized liquid–solid contact area and the fluorocarbon layer deposited during the DRIE process, this silicon mold shows super-hydrophobicity, and its surface energy is sharply reduced to be ultralow. Consequently, the intermolecular force between PDMS and ultra-low-surface-energy mold is significantly weakened, and the undesirable bonding of PDMS to silicon substrate that would otherwise occur at high temperature was prevented, even for temperature higher than 180 °C. Therefore, micro-/nanohierarchical structures could be easily replicated from silicon substrate to PDMS film without surfactant coating. A major advantage of this method is that the precision of the pattern replication is improved, as surfactant coating is not required.

3.2 Micro-/Nanohierarchical Structures Based on PDMS

3.2.1 Brief Introduction of PDMS

PDMS has attracted much attention in a wide range of fields due to its unique properties, such as transparency, flexibility, biocompatibility, and high yield. Especially, the super-hydrophobic PDMS has shown potential applications in self-cleaning devices and microfluidic systems. Generally, the preparation of PDMS

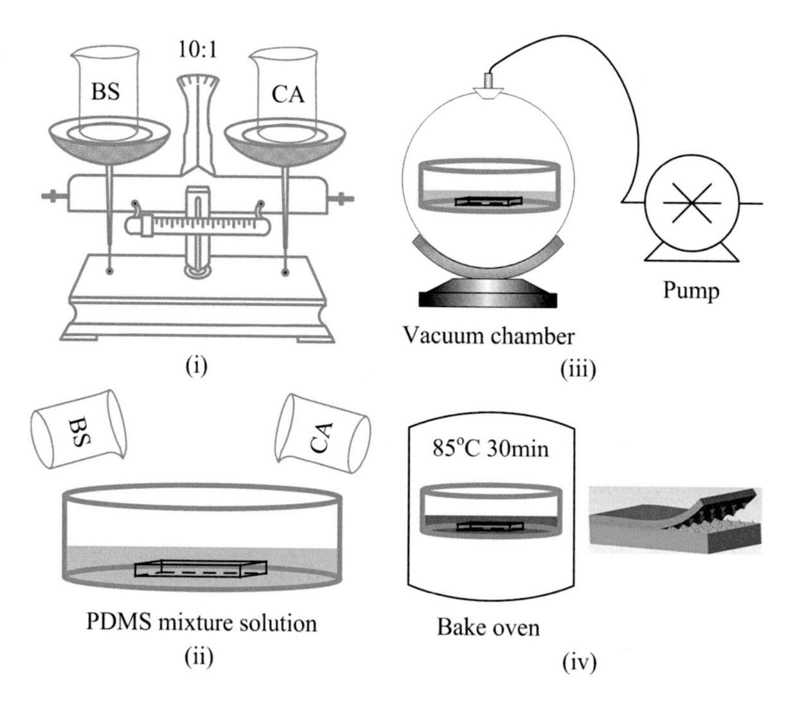

Fig. 3.3 The fabrication process flow of PDMS film

film from PDMS solution is polymerization reaction (i.e., cross-linking process). Commercial PDMS product (Sylgard 184, Dow Corning Corp.) was used for structure replication. The fabrication of micro-/nanohierarchical structure on PDMS can be briefly summarized as follows and shown in Fig. 3.3.

(i) Base solution and curing agent were mixed with the weight ratio of 10:1.
(ii) The silicon mold was put at the bottom of a container, and then, the PDMS mixture solution was poured into the container to cover the silicon mold.
(iii) Sequentially, the container was vacuumized at ~1×10^{-2} MPa for 15 min in a vacuum chamber to remove bubbles in the interface between substrate and PDMS mixed solution.
(iv) Finally, the samples were heated at high temperature from 60 to 95 °C for 30 min, in order to cure the liquid PDMS into a solid PDMS membrane. Then, the solid PDMS film was simply peeled off from the silicon mold.

There are two key points during the replication process. First, as the replication mold, the prepatterned silicon substrate is placed at the bottom of container, and more importantly, the micro-/nanohierarchical structures should be placed upward. Second, the bubbles at the interface between PDMS solution and silicon mold must be removed completely by using the vacuuming process.

3.2.2 *Single-Step Fabrication of Micro-/Nanohierarchical Structures on PDMS*

Figure 3.4 illustrates the process flow of fabricating micro-/nanohierarchical structures on PDMS, which can be summarized as follows: (I) fabricating periodic microstructures on silicon substrate by using traditional microfabrication techniques; (II–IV) forming nanostructures atop microstructures by using the improved DRIE process and then obtaining Si-based micro-/nanohierarchical structures; (V) depositing a nanoscale fluorocarbon polymer later atop silicon mold; and (VI–VII) replicating micro-/nanohierarchical structures from silicon mold to PDMS film. It is worth mentioning that the process flow from II to V is a single-step process, i.e., fully optimized DRIE process shown in Fig. 3.2. The process parameters of this fully optimized DRIE process are shown in Table 3.1.

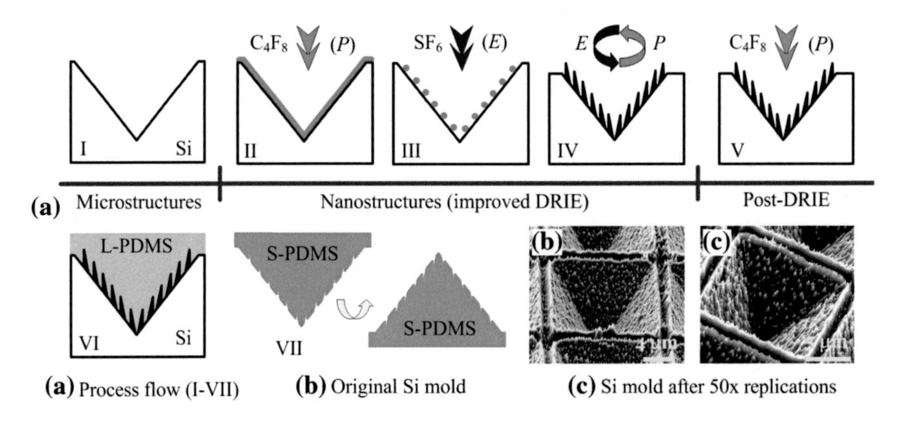

Fig. 3.4 **a** Schematic view of producing (I–V) ultra-low-surface-energy silicon substrate with micro-/nanohierarchical structures and (VI–VII) pattern transfer to PDMS membrane by replication process. SEM images of Si substrate after PDMS replication for **b** 0 time and **c** ~50 times, respectively

Table 3.1 The process parameters of the improved DRIE process and the post-DRIE process[a]

	Gas flow (sccm)		RF power (W)	Platen power (W)	Pressure (mTorr)	Time ratio of E/P[b]	Cycle of E/P[b]
	SF_6	C_4F_8					
Improved DRIE	30	50	825	10.5	12	6 s:6 s	70
Post-DRIE	0	50	825	0	12	6 s:6 s	1, 2, 4, 10

[a]The improved DRIE process and the following post-DRIE process actually belong to one recipe of the ICP equipment. In other words, we set up the ICP equipment using only one recipe, which consists of the improved DRIE process and the following post-DRIE process. The silicon wafer is placed into the reactive chamber for one time only, and then, the silicon mold with micro-/nanohierarchical structures is obtained

[b]E/P—etching step and passivation step

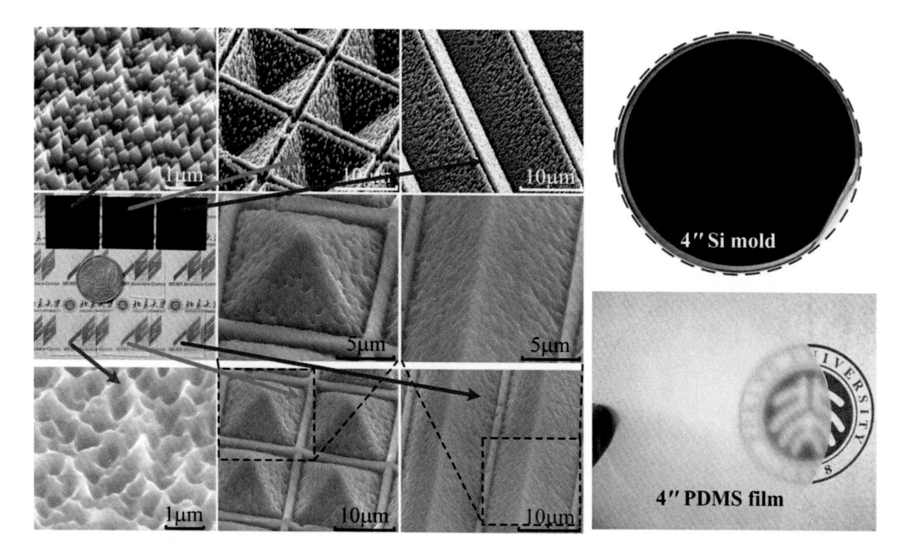

Fig. 3.5 Photographs and SEM images of surface-micro-/nanostructured PDMS membranes and their replica molds, including 4-in. samples

The SEM images of micro-/nanohierarchical structures on silicon mold before and after 50× replication processes are compared and shown in Fig. 3.4b, c. After dozens of replication processes, the surface of silicon mold almost preserved original state without visible damage, and both periodic microstructures and high-dense nanostructures kept highly consistent. And this observation demonstrates that the interface force between PDMS and silicon mold can be almost ignored due to the ultra-low surface energy resulting from the fully optimized DRIE process, and thus, peeling-off steps have no effect on the mold surface.

In order to deeply study the present micro-/nanointegrated fabrication technique for flexible materials, three types of surface geometries were designed, namely high-dense nanostructures, pyramid-shaped hierarchical structures, and groove-shaped hierarchical structures, as shown in Fig. 3.5. Here, in order to simplify the description, PDMS film with pyramid-shaped micro-/nanohierarchical structures is abbreviated as MNHS-P PDMS, while PDMS film with groove-shaped hierarchical structures is abbreviated as MNHS-G PDMS.

Similar to the Si-based micro-/nanointegrated fabrication technique present in Chap. 2, the micro-/nanointegrated fabrication technique for flexible materials present here also shows the advantages of mass fabrication and cost-efficient, which can be used to fabricate wafer-level samples directly. This was demonstrated by fabricating 4-in. samples shown in Fig. 3.5. According to the SEM images shown in Fig. 3.5, micro-/nanohierarchical structures, including both periodic microstructures and high-dense nanostructures, were successfully transferred from silicon mold to PDMS film. Even the tilted surfaces of microstructures were completely covered by nanostructures (i.e., nanoporous). Additionally, all PDMS samples show a remarkable transparency.

3.2.3 Key Factors of Single-Step Replication Process

When replicating micro-/nanohierarchical structures from silicon mold to PDMS surface, there are two key factors affecting the pattern transfer result, including baking temperature and time. According to the explanation in Sect. 3.2.1, the preparation of PDMS film can be simply summarized as baking the mixture solution of PDMS base solution and curing agent. Thus, when the weight ratio of base solution and curing agent is fixed as 10:1, the baking temperature becomes the core factor affecting the physical property of PDMS film. Within a certain range, the higher the baking temperature, the greater the mechanical strength of the PDMS film. Thus, the MNHS PDMS structures baked at higher temperature are stronger and less likely to be destroyed when they are peeled off from the mold, which results in better pattern transfer.

According to the experimental results, this relationship has been demonstrated in the baking temperature range from 60 to 85 °C. The SEM and AFM images of surface-nanostructured PDMS samples illustrate that the pattern transfer precision increases as the baking temperature increases from 60 to 85 °C, as shown in Figs. 3.6a and 3.7.

The surface-nanostructured PDMS film baked at 60 °C is smoother than that baked at 85 °C. In Fig. 3.7, the average roughness (i.e., R_a) of surface-nanostructured PDMS sample increases during the baking temperature range from 60 to 85 °C.

$$R_{a|60^\circ C}(=76.2) < R_{a|80^\circ C}(=87.6) < R_{a|85^\circ C}(=93.7) \tag{3.1}$$

$$R_a = \frac{1}{N}\sum_{j=1}^{N}|Z_j| \tag{3.2}$$

in which N is the number of nanostructures and Z_j is the nanostructure height of number j.

In contrast, when the baking temperature is higher than 85 °C, the pattern transfer precision decreases with the further temperature increase.

$$R_{a|60^\circ C}(=76.2) < R_{a|95^\circ C}(=78.4) < R_{a|80^\circ C}(=87.6) \tag{3.3}$$

We believe that the difference between baking temperature and room temperature affects the PDMS replication due to thermal expansion and contraction. The larger the temperature difference, the greater the thermal contraction. Therefore, the baking temperature and its difference in room temperature have opposite effects on the pattern transfer. According to the experiments shown in Fig. 3.7, the balance point occurs at the temperature of 85 °C and the maximum roughness increased up to 93.7. The static contact angle (CA) of surface-nanostructured PDMS films has also been investigated and shows the same trend with a peak point at 85 °C, as shown in Figs. 3.9 and 3.10 (black lines).

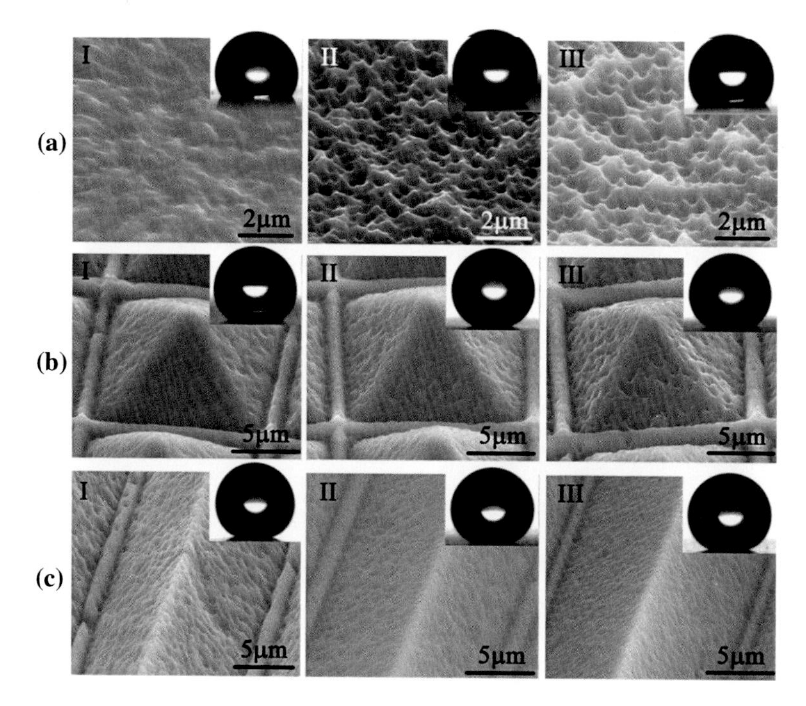

Fig. 3.6 SEM images and contact angle (CA) photographs of PDMS samples fabricated at (I) 60 °C, (II) 85 °C, and (III) 95 °C. (*upper* surface-nanostructured PDMS films; *middle* MNHS PDMS films with pyramid arrays; *bottom* MNHS PDMS films with groove arrays). **a** Nanoporous, **b** pyramid-shaped MNHS, **c** groove-shaped MNHS

The effect of baking time has also been studied. The PDMS cross-linking procedure (i.e., solidification) is actually the combination of small polymeric chains. The PDMS solution gradually transforms to the solid film as the baking time increases. According to our experiment, the cross-linking procedure of PDMS finished after 30 min of baking. As is shown in Fig. 3.11, the CA increases to the maximum within 30 min and keeps constant after complete curing. This phenomenon will be further studied in Sect. 3.2.4.

3.2.4 *Effect of Process Parameters on Surface Properties of PDMS*

Increasing surface roughness has been demonstrated to be an effective method to enhance the intrinsic wettability of material. For example, if the material is water-repellent, its hydrophobicity can be strengthened by increasing surface roughness. As is known, the intrinsic wettability of PDMS is hydrophobic, and the CA of flat PDMS surface is about 110.6° shown in Fig. 3.8.

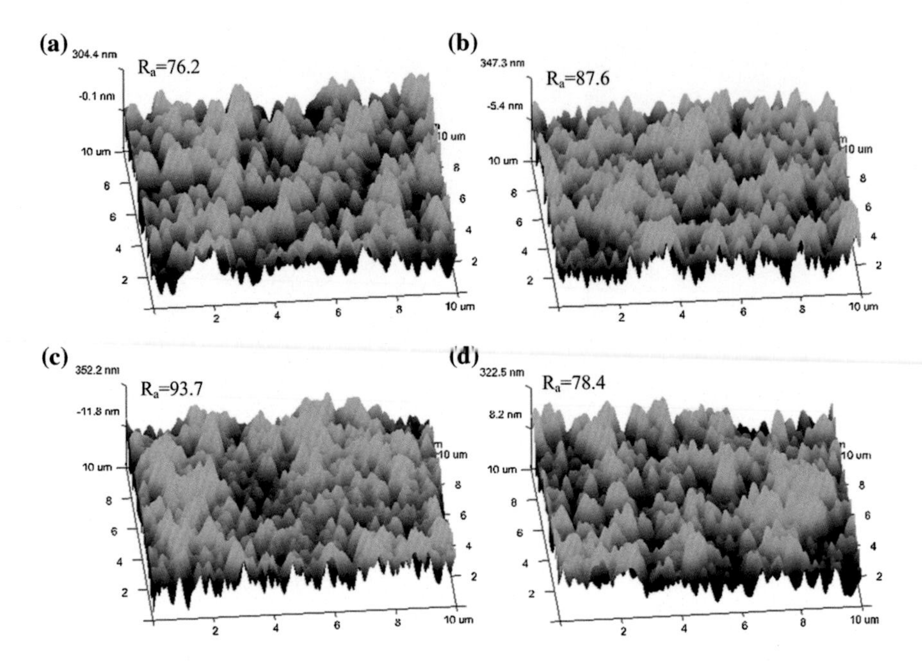

Fig. 3.7 The atomic force microscopy (AFM) images of surface-nanostructured PDMS membrane fabricated at **a** 60 °C, **b** 80 °C, **c** 85 °C, and **d** 95 °C. (R_a—average roughness)

Therefore, increasing surface roughness by micro-/nanostructures can enhance the hydrophobicity of PDMS and enlarge its CA. In the meantime, the micro-/nanostructures on PDMS film depend on the pattern transfer precision of replication process, and higher precision causes larger roughness. Therefore, the CA

Fig. 3.8 The static contact angle of flat PDMS film (CA = 110.6°)

values of fabricated PDMS samples can be used to estimate the pattern transfer precision of replication process. The static CA and the stability of super-hydrophobicity of micro/nano dual-scale PDMS surfaces were measured by an OCA 20 video-based CA meter (DataPhysics Instruments GmbH). The physical mechanism of the effect of baking temperature and time on pattern transfer was analyzed by using an atomic force microscope (Dimension ICON, Bruker Corp.) and a scanning electron microscope (Quanta 600F, FEI Co.).

Figure 3.9 shows the effect of baking temperature on CA of surface-textured PDMS film, in which two obvious trends can be observed. First, the CAs of surface-nanostructured PDMS and MNHS-P PDMS became larger firstly and subsequently lowered as the baking temperature increased continuously. During the temperature range of 60–85 °C, CAs of surface-nanostructured PDMS increased from 121.1° to 136.8°, while CAs of MNHS-P PDMS increased from 135.4° to 148.8°. However, when the baking temperature increased further from 85 to 95 °C, CAs of surface-nanostructured PDMS decreased from 136.8° to 128.2°, while CAs of MNHS-P PDMS increased from 148.8° to 139.8°. This changing trend demonstrates the analysis in Sect. 3.2.3, which results from the balance point of opposite effects of the baking temperature and its temperature difference in room temperature on the pattern transfer. Second, the CA curve of MNHS-G PDMS shows completely opposite trend, which results from the interaction effect of hierarchical multiscale structures. This new phenomenon will be explained in detail in Sect. 3.4 shown as "The Interaction of Multiscale Structures on Flexible Materials".

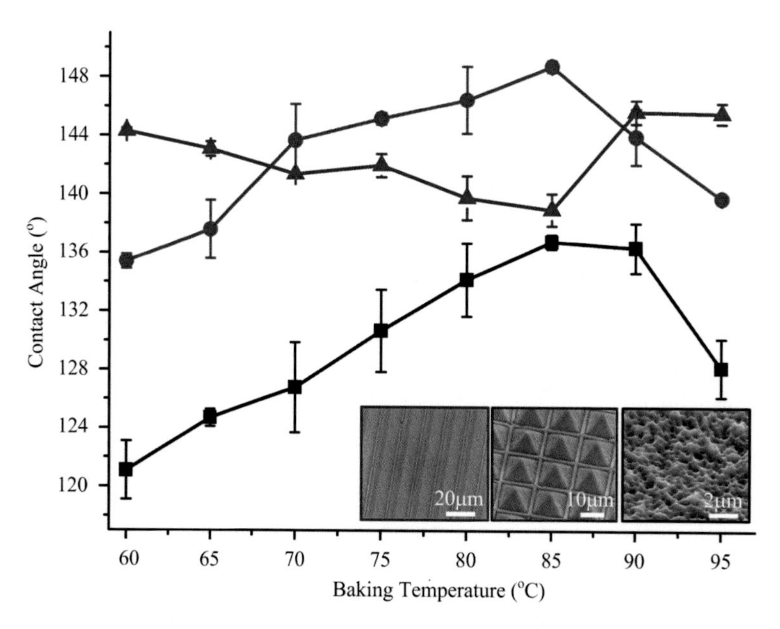

Fig. 3.9 The effect of baking temperature on the static contact angle of surface-micro-/nanostructured PDMS membrane (water droplet volume = 2 μL)

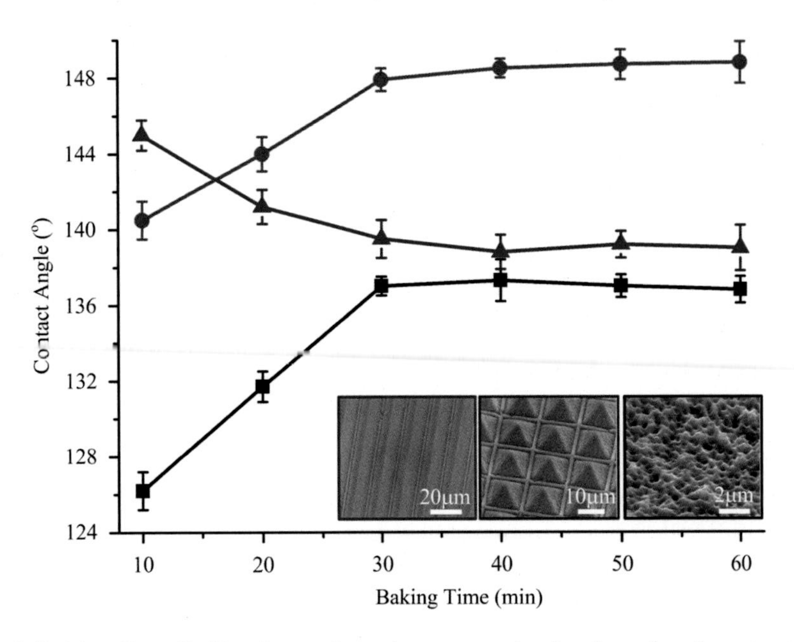

Fig. 3.10 The effect of baking time on the static contact angle of surface-micro-/nanostructured PDMS membrane (water droplet volume = 2 μL)

Besides the baking temperature, the effect of baking time on CAs of fabricated samples was investigated, as shown in Fig. 3.10. Similar to Fig. 3.9, two opposite trends were observed. The CA curves of surface-nanostructured PDMS and MNHS-P PDMS rose firstly and declined subsequently as the baking time increased continuously, while the CA curve of MNHS-G PDMS showed a converse trend. However, all of them possessed the same turning point at the baking time of 30 min. After baked for half an hour, CAs of all samples kept constant, which indicates the completeness of cross-linking procedure of PDMS mixture solution.

Thus, based on this single-step micro-/nanointegrated fabrication technique, we successfully realized micro-/nanohierarchical structures on PDMS with attractive hydrophobicity. Additionally, these PDMS films with micro-/nanohierarchical structures possess attractive potential to be utilized as functional materials. For pyramid-shaped micro-/nanohierarchical structures, they can be used as a light absorption coating layer. Pyramid arrays were used to enhance light absorption and minimize reflectance for decades due to the multireflectance by inclined sidewalls of the pyramid [3–5]. Here, the MNHS-P PDMS films show more potential for future applications as an effective light absorption coating layer for optical microsystems due to the high-efficiency light trap by the combination of microscale pyramid arrays and nanoscale holes.

In the meantime, MNHS-G PDMS films show an anisotropic wettability (Fig. 3.11), which makes it an attractive material used for realizing the directional

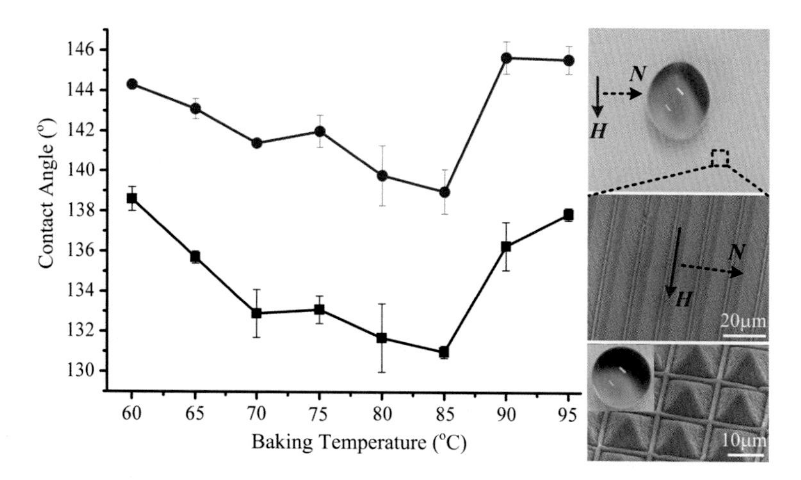

Fig. 3.11 Contact angles of MNHS-G PDMS film in horizontal (*H*) and normal (*N*) directions

transport of liquid droplets. The water droplet with blue ink showed a profile of round on the surface of MNHS-P PDMS, while a profile of ellipse on the surface of MNHS-G PDMS. Thus, MNHS-P PDMS films show an isotropic wettability, and in other words, the CAs keep constant, regardless of the observation direction. In contrast, MNHS-G PDMS films show an anisotropic wettability, and different CAs are obtained depending on the observation direction. In detail, CAs in "*H*" direction were higher than those in "*N*" direction by about 10°, as shown in Fig. 3.11. Here, "*H*" and "*N*" directions mean the horizontal direction and the normal direction, respectively, which are parallel with and normal to grooves, correspondingly. Therefore, it is easier for water droplets to move along the "*H*" direction, which is very useful to realize the directional transport of liquid droplets.

3.3 Micro-/Nanohierarchical Structures Based on Parylene-C

Besides PDMS, parylene-C is another important polymer that is stronger, biocompatible, waterproof, and transparent. Thus, it is widely used as an encapsulation layer of micro-/nanodevice, especially for implantable biomedical microsystems and wearable devices [6–8]. Patterning the surface is a useful method to functionalize parylene-C films and is one of the most important research fields for parylene-C. Several techniques, such as photolithography and replication process, were developed to pattern the parylene-C surface [9, 10]. However, it is still hard to realize micro-/nanohierarchical structures by using the conventional techniques due to the minimum width limitation of photolithography and the complicated multiple steps of replication process. Here, the micro-/nanointegrated fabrication technique for flexible materials presented above is also explored with parylene-C, and a single-step replication process based on the ultra-low-surface-energy silicon mold is successfully proposed.

3.3.1 Fabrication and Method

Before introducing the fabrication of MNHS parylene-C, it is necessary to briefly explain the fabrication of parylene-C. The parylene-C thin film is formed by vapor deposition process generally including three main steps. Firstly, the parylene-C precursor (i.e., dimer) in the form of small pellets is placed into the vaporizer, and then, it will be vaporized and heated to form the dimeric gas (i.e., dimer molecule) under vacuum. Secondly, this dimeric gas is heated in the pyrolysis furnace and then pyrolized to cleave the dimer into its monomeric form. Finally, this monomer polymerizes and forms a thin film atop the substrate placed in the coating chamber. Figure 3.12 shows the deposition process of parylene-C membrane.

Figure 3.13 shows the fabrication process flow of MNHS parylene-C. Compared with the fabrication process of MNHS PDMS shown in Fig. 3.4, the fabrication process of MNHS parylene-C is similar and also based on the single-step replication from ultra-low-surface-energy silicon mold. The key points behind this single-step replication process can be summarized as follows: first, the controllable fabrication of ultra-low-surface-energy silicon mold and second, the pattern replication of parylene-C film. Figure 3.14 shows the photographs and SEM images of fabricated surface-textured parylene-C films. The success of extending this single-step replication process from PDMS to parylene-C enlarges its application range and makes this micro-/nanointegrated fabrication technique suitable for flexible materials.

Two types of micro-/nanohierarchical structures were fabricated, including pyramid shape (i.e., MNHS-P) and groove shape (i.e., MNHS-G) that are shown in

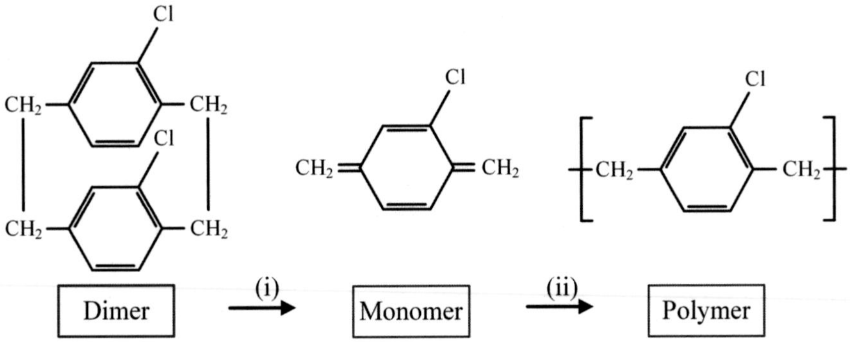

Fig. 3.12 Schematic view of the deposition process of parylene-C membrane

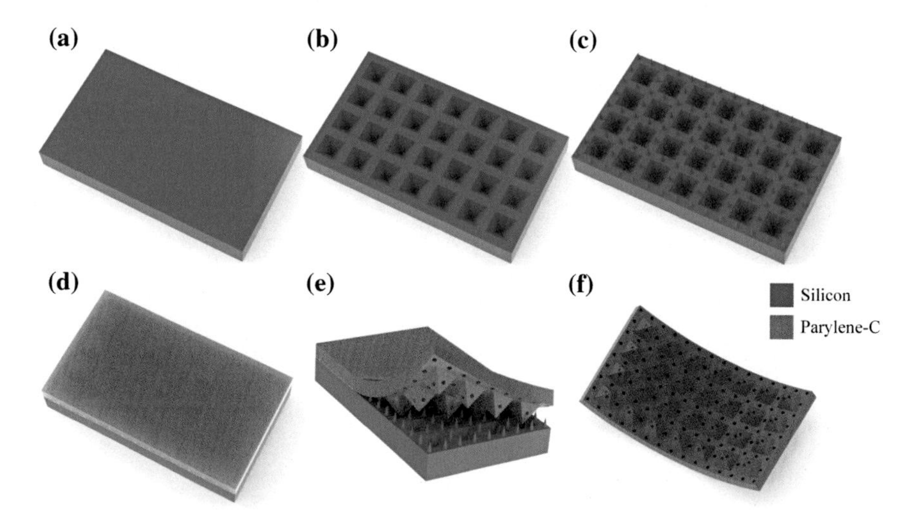

Fig. 3.13 The fabrication process flow of MNHS parylene-C film

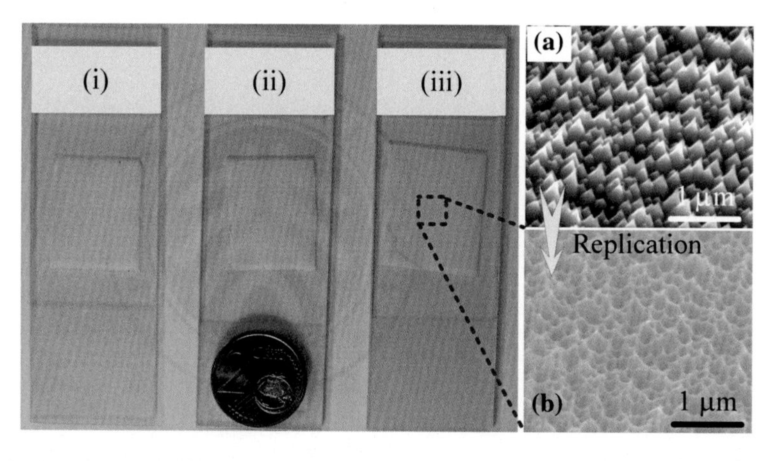

Fig. 3.14 The photographs of surface-micro-/nanostructured parylene-C films: (i) groove-shaped MNHS, (ii) pyramid-shaped MNHS, and (iii) nanoporous

Figs. 3.15 and 3.16, respectively. Comparing the SEM images of silicon molds and MNHS parylene-C films, we found that patterns transferred from silicon molds to surfaces of parylene-C were perfect. The high-magnification views clearly show that even the tiny nanoscale cones were replicated very well to form nanoscale porous on the surface of parylene-C film. Compared with the fabricated MNHS PDMS shown in Fig. 3.6, the fabricated MNHS parylene-C films show better pattern transfer results without visible distortions. It is believed that the remarkable mechanical properties of parylene-C play an important role, which prevent

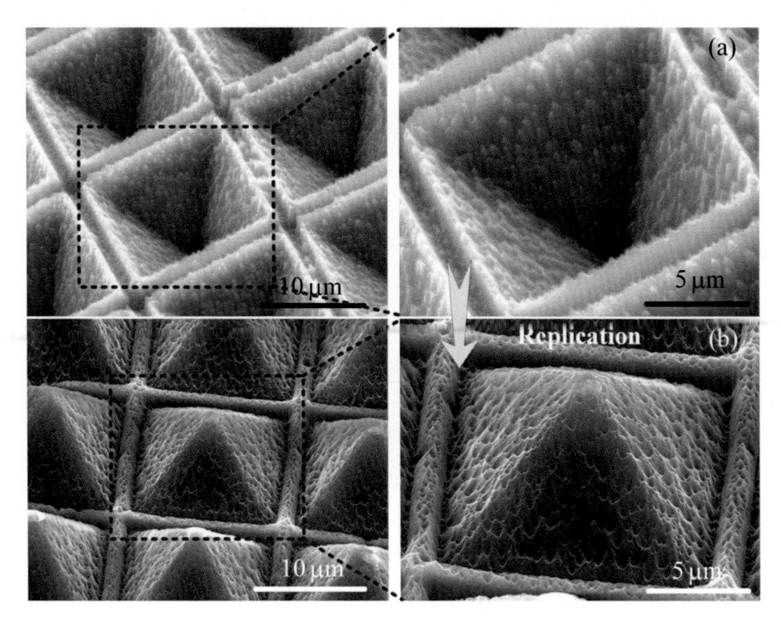

Fig. 3.15 SEM images of **a** MNHS-P silicon mold and **b** the fabricated MNHS-P parylene-C film

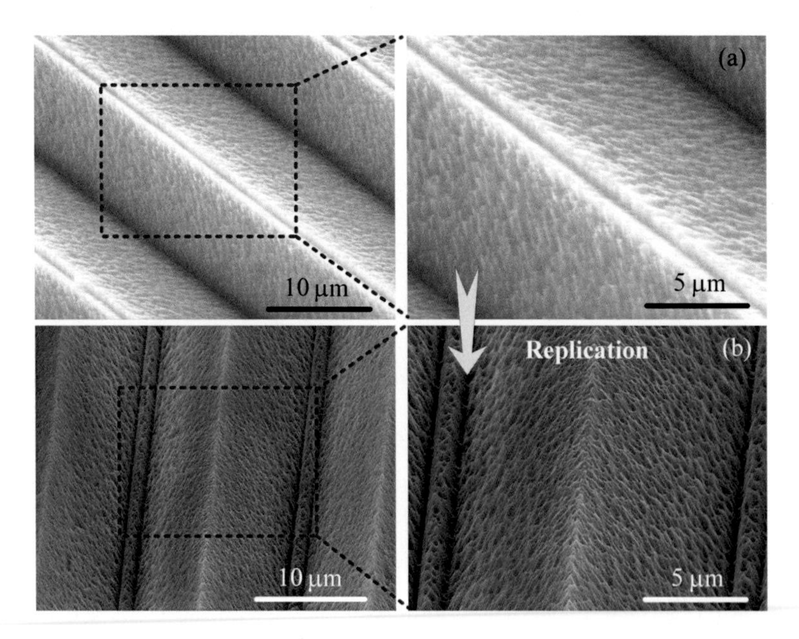

Fig. 3.16 SEM images of **a** MNHS-G silicon mold and **b** the fabricated MNHS-G parylene-C film

the unwanted damage to micro-/nanohierarchical structures during the peeling-off procedure. Additionally, compared with PDMS samples of hundred micrometer thickness, parylene-C films with a few micrometer thicknesses were fabricated more easily, such as the 13-μm-thick parylene-C samples shown in Fig. 3.14.

3.3.2 The Properties of Fabricated MNHS Parylene-C Films

The MNHS parylene-C films show enhanced hydrophobicity due to the combination of microstructures and nanostructures, which is similar to MNHS PDMS shown in Sect. 3.2. Figure 3.17 shows the test results of CAs of 2 μL water droplet on the surfaces of parylene-C films. As shown in Fig. 3.17a, the original parylene-C film with flat surface shows a weak hydrophobicity and its CA is about 91°. After micro-/nanostructuring surfaces of parylene-C films, the CAs of fabricated samples were enhanced significantly. The CA of surface-nanostructured sample increased up to 125°, while the CAs of the samples with micro-/nanohierarchical structures were enhanced further to more than 130°. The CAs of MNHS-P and MNHS-G parylene-C films increased to 133° and 135°, respectively. Generally, the CA of parylene-C film was enhanced 1.48-fold by micro-/nanohierarchical structures in comparison with the flat surface.

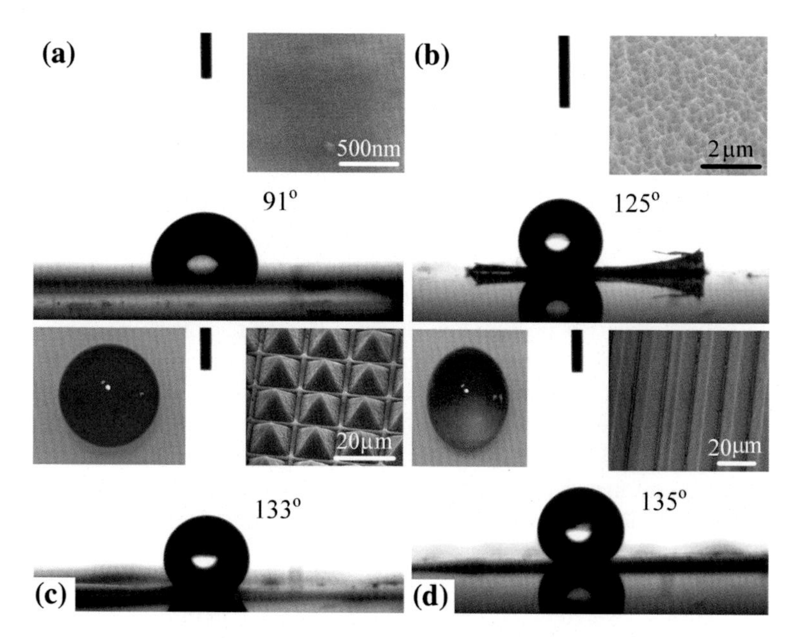

Fig. 3.17 The CAs of fabricated parylene-C films: **a** flat surface, **b** nanoporous surface, **c** MNHS-P surface, and **d** MNHS-G surface

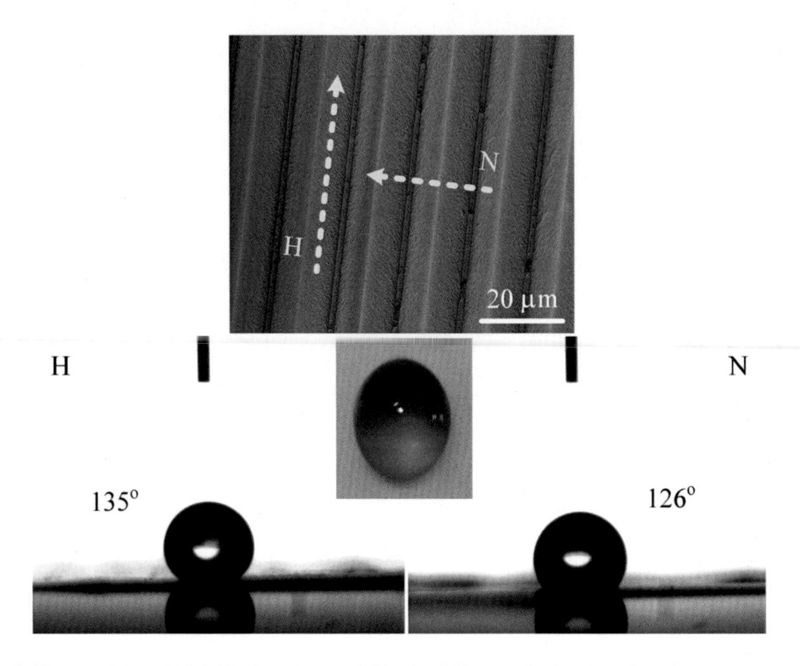

Fig. 3.18 The CAs of MNHS-G parylene-C film in different observation directions

In addition, the anisotropic wettability was observed on the surface of MNHS-G parylene-C film resulting from the specific geometry of grooves, which is similar to MNHS-G PDMS. As shown in Fig. 3.18, the CA of MNHS-G parylene-C film in parallel direction (i.e., *H* direction) is 135°, while the CA in normal direction (i.e., *N* direction) is 126°. This anisotropic wettability shows the attractive potential of MNHS-G parylene-C film in realizing the directional transport of liquid droplets.

As a universal detection technology, the Raman spectroscopy has been widely used in various fields, especially for biomedical applications, because of its attractive features of low cost, rapid response, non-destruction, and so on [11, 12]. However, natural materials possess low-sensitive Raman response. In other words, the concentration threshold of target detection is very high. In order to overcome this drawback, the surface-enhanced Raman scattering (SERS) technique is developed for the trace detection even at the molecular level, which is widely used in chemical detection and biomedical analysis [13–15]. Increasing the surface roughness of noble metal materials is one of the most important methods to strengthen the intensity of the Raman scattering spectrum [16]. In this work, due to the combination of periodic microstructures and high-density nanostructures, the fabricated parylene-C films show remarkable SERS properties. The Raman spectra of fabricated samples covered by 800 Å gold layer are shown in Fig. 3.19.

Compared with the flat sample, the surface-textured parylene-C film shows outstanding SERS properties, achieving a maximum enhancement factor

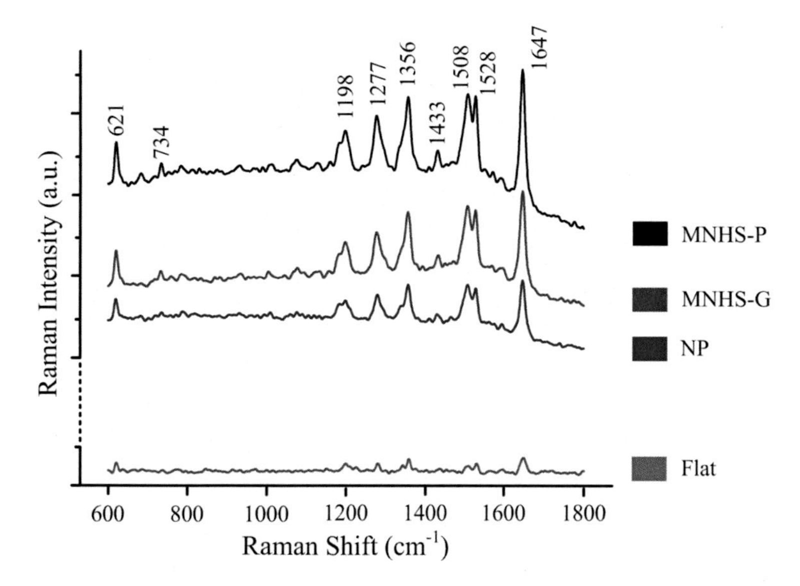

Fig. 3.19 The surface-textured parylene-C films covered by 800 Å Au show remarkable surface-enhanced Raman scattering (i.e., SERS) property, which indicates the highest enhancement factor of 2.3×10^4 compared with the flat one (MNHS-P—pyramid-shaped micro-/nanohierarchical structures; MNHS-G—groove-shaped micro-/nanohierarchical structures; NP—nanoporous)

of 2.3×10^4 with the pyramid-shaped micro-/nanohierarchical structures. The micro-/nanohierarchical structures further enhance the Raman intensity than either microstructures or nanostructures. In other words, the parylene-C film with micro-/nanohierarchical structures is more sensitive and possesses higher resolution. It was observed that the nanostructure is the dominant contributor to the SERS properties due to the similar Raman intensity between micro-/nanohierarchical structures and nanostructures. Additionally, the fabricated parylene-C films show the reusable property and the micro-/nanohierarchical structures recover the original state by a simple cleaning step of acetone and ethanol, as shown in Fig. 3.20.

3.4 The Interaction of Multiscale Structures on Flexible Materials

During the fabrication of micro-/nanohierarchical structures on flexible materials, a new phenomenon was observed, as shown in Figs. 3.9 and 3.10. The trends of CAs of MNHS-P and MNHS-G PDMS films changing with the baking temperature are opposite. It is very hard to find a suitable way to explain this phenomenon by using the traditional theory of fluid dynamics. Herein, a reasonable explanation is given by finite element analysis (FEA).

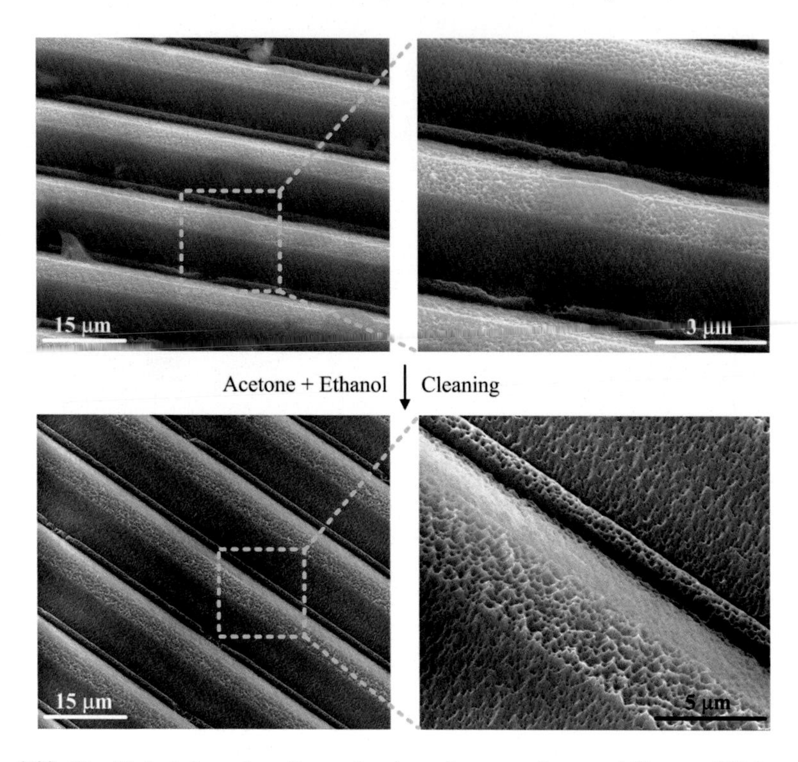

Fig. 3.20 The fabricated parylene-C samples showed an attractive reusability as a SERS surface by a simple cleaning step of acetone and ethanol

In Sect. 2.3, we studied the interaction of multiscale structures on silicon substrate. Similarly, the interaction of multiscale structures during the micro-/nanointegrated fabrication procedure of flexible materials was also investigated here and believed to be the key point to induce this observed phenomenon. Although many factors correlate with the pattern transfer result, such as baking time and temperature, the step of peeling off directly affects the replication process. Taking PDMS as example, we used COMSOL software to analyze the stress distribution on the surface.

As shown in Fig. 3.21, the stress distributions on the surfaces of MNHS-P and MNHS-G PDMS films are different. For the pyramid-shaped (MNHS-P) samples, the stress only distributes at the bottom of the microscale pyramid; thus, the nanoscale holes experience almost zero stress and will keep their original profile during the peeling-off step. In contrast, for the groove-shaped (MNHS-G) samples, the stress distribution concentrates at the bottom of the microscale groove as well as in the nanoscale holes. And the stress distributed in nanoholes will enlarge the aspect ratio of the nanoscale holes during the peeling-off process. Logically, the softer the PDMS film is, the larger the stress effect on the aspect ratio of nanoholes will be. As is known, the Young's modulus of the PDMS film will become larger when the baking temperature is higher. Therefore, when a lower

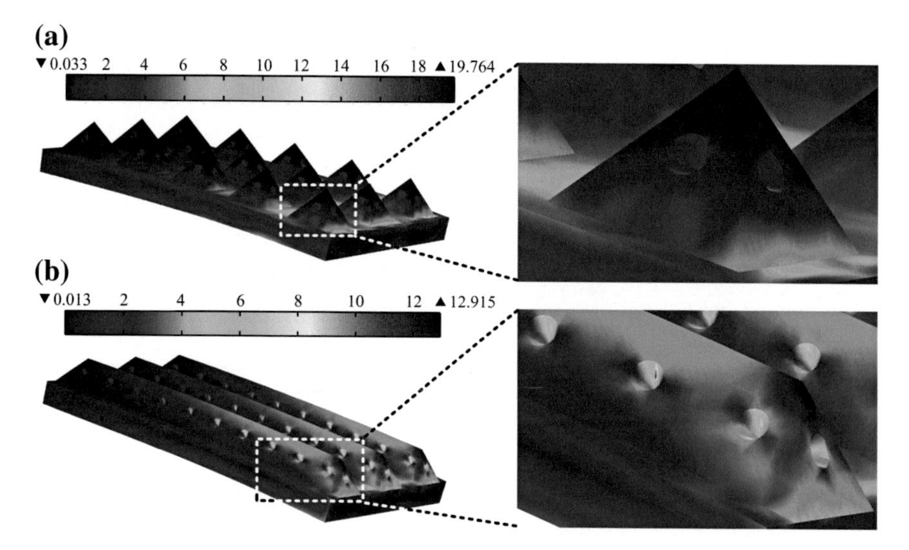

Fig. 3.21 The FEA study of the stress distribution on MNHS PDMS surfaces during the peeling-off procedure from silicon molds

temperature or a shorter time is applied to bake the MNHS-G PDMS, the stress effect will make the aspect ratio of nanoholes larger and subsequently increase the CA. Thus, the interaction of hierarchical multiscale structures (i.e., the effect of the microstructure morphology on the nanostructure formation) explains the opposite curve trends of MNHS-P and MNHS-G PDMS shown in Figs. 3.9 and 3.10.

3.5 The Surface Modification Based on Post-DRIE Process

In Sect. 3.1.2, we demonstrate the post-DRIE process as an efficient approach to reduce the surface energy of silicon substrate by coating a thin fluorocarbon polymer. Herein, in order to further strengthen the water-repellent property of PDMS to be super-hydrophobicity, this post-DRIE process is also employed to modify the surface of PDMS.

3.5.1 Fluorocarbon Plasma Treatment Based on Post-DRIE Process

The process parameters of post-DRIE are shown in Table 3.1, and C_4F_8 gas was used as the passivation gas. 1-cycle, 2-cycle, 4-cycle, and 10-cycle recipes were studied and shown in Fig. 3.22. The time of one cycle is 12 s without the etching gas, and in other words, the effective process is the passivation step that lasts 6 s.

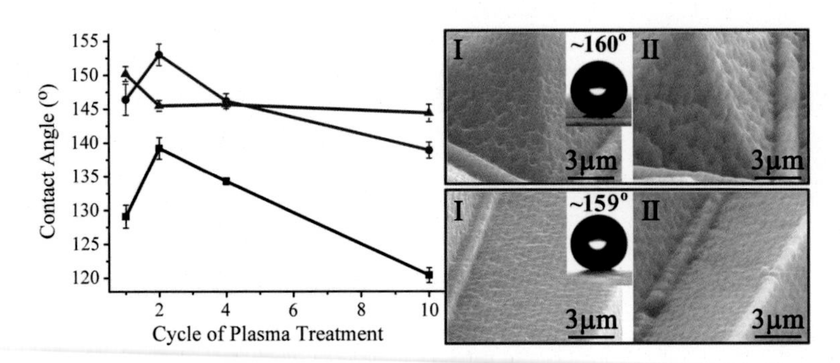

Fig. 3.22 The effect of the post-DRIE process (i.e., C_4F_8 plasma treatment) cycle on CA of surface-micro-/nanostructured PDMS (3 μL) and SEM images of MNHS PDMS film after (I) 2 cycles and (II) 10 cycles of plasma treatment, respectively

According to the experimental results, the CA of PDMS film increases first, while it decreases later as the cycle of post-DRIE increases. When the cycle increased from 1 to 2, the CA increased sharply resulting from the fluorocarbon polymer significantly reducing the surface energy. However, when the cycle further increased from 2 to 10, the thickness of fluorocarbon polymer further increased too and covered the surface structures of PDMS. Consequently, the surface roughness of PDMS decreased resulting in the decrease of CA. The above analysis was demonstrated by SEM images of fabricated samples shown in Fig. 3.22. Compared with the 2-cycle sample, the nanostructures on 10-cycle sample were covered by a thicker polymer that reduces the surface roughness. The largest CA was obtained when the cycle of post-DRIE process was set as two. With the optimized structural parameters (i.e., 13 μm width and 2 μm space), the maximum CA reaches to about 160°, while the rolling angle reduces to below 5°.

The hydrophobic stability of PDMS samples treated by the post-DRIE process is verified by squeezing test (droplet volume of 3 μL) and impact test (droplet volume of 4.5 μL). In Fig. 3.23a, during the vertical movement process of the platform, although the water droplet is pushed, squeezed, and pulled sequentially, it still remains on the injector tip. In Fig. 3.23b, when the water droplet drops from the injector tip and impacts the sample surface, it bounces several times along the 10° inclined surface. Furthermore, the above experiments were carried out two weeks later after the post-DRIE process, which further verifies the hydrophobic reliability of fabricated PDMS samples.

3.5.2 The Mechanism of Enhancing Hydrophobicity by Using Post-DRIE Process

To understand the chemical mechanism behind the effect of C_4F_8 plasma treatment on the PDMS surface energy, we studied the interaction of fluorocarbon

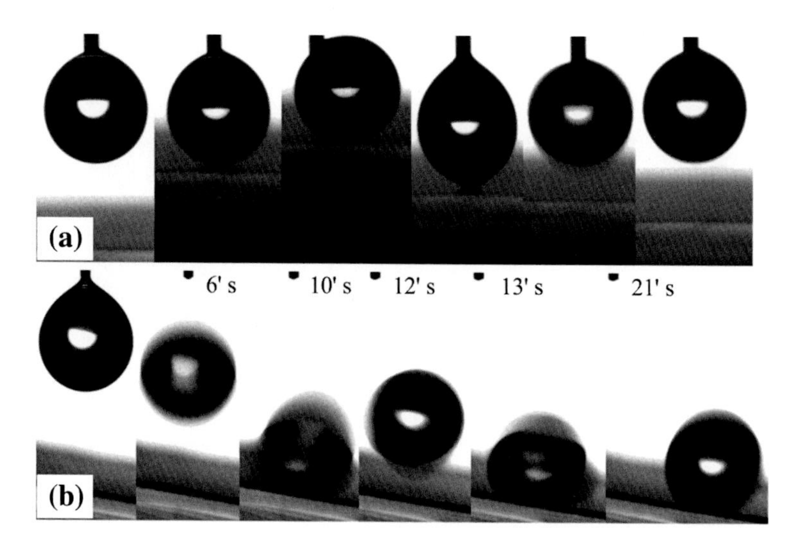

Fig. 3.23 The super-hydrophobic stability test of PDMS samples treated by the post-DRIE process: **a** squeezing test and **b** impact test

polymer deposited by C_4F_8 plasma and PDMS with H_2O by optimizing two model complexes, $CF_3(CF_2)_4CF_3{\cdot}H_2O$ and $(CH_3)_3SiOSi(CH_3)_2OSi(CH_3)_3{\cdot}H_2O$, respectively, at the DFT level. DFT calculations were performed on $CF_3(CF_2)_4CF_3$ and $(CH_3)_3SiOSi(CH_3)_2OSi(CH_3)_3$ and their hydrated complexes $CF_3(CF_2)_4CF_3{\cdot}H_2O$ and $(CH_3)_3SiOSi(CH_3)_2OSi(CH_3)_3{\cdot}H_2O$ using the generalized gradient approximation (GGA) with the PBE exchange–correlation functional [17] as implemented in the Amsterdam Density Functional (ADF 2010.02) program [18–20]. The Slater-type basis sets with the quality of double-ζ plus one polarization function (DZP) [21] were used, with the frozen core approximation applied to the inner shells $[1s^2–2p^6]$ for Si and $[1s^2]$ for C, O, and F. The scalar relativistic (SR) effects were taken into account by the zero-order regular approximation (ZORA) [22]. Geometries were fully optimized, and vibration frequencies were calculated to verify the local minima on the energy surface at the SR-ZORA level.

For convenience of statement, we denote the two model complexes by $MCF{\cdot}H_2O$ and $MSiO{\cdot}H_2O$, respectively. The two optimized stable complexes are shown in Fig. 3.24. Obviously, in the $MCF{\cdot}H_2O$ complex, water binds to one terminal F atom and forms a hydrogen bond in a monodentate fashion with a H···F bond length of 2.45 Å, while in $MSiO{\cdot}H_2O$, water binds to one terminal O atom and also forms a hydrogen bond with a H···O bond length of 1.89 Å. The hydrogen bond interaction in $MSiO{\cdot}H_2O$ is higher than that in $MCF{\cdot}H_2O$ by about 7 kcal/mol; i.e., the H_2O is more attracted by PDMS than the fluorocarbon polymer, due to the stronger electrostatic attraction in the former. The Mulliken charge analysis in Fig. 3.24 shows that the O atom carries more negative charge than the F atom, i.e., −0.97 versus −0.45, and vice versa for the two H atoms involved in the hydrogen bonds.

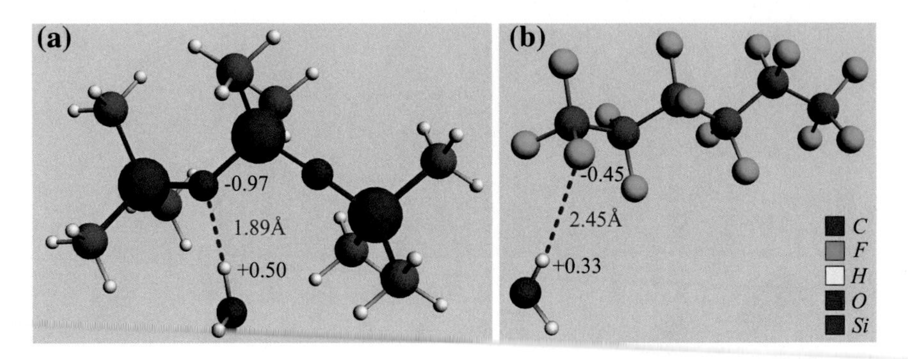

Fig. 3.24 The study of the interaction between water molecule and PDMS samples **a** before and **b** after the post-DRIE process (i.e., C_4F_8 plasma treatment) via optimization of model complexes $(CH_3)_3SiOSi(CH_3)_2OSi(CH_3)_3 \cdot H_2O$ and $CF_3(CF_2)_4CF_3 \cdot H_2O$ at the density functional theory level. In these two stable complexes, the bond lengths of hydrogen bonds, i.e., H···O and H···F, and the Mulliken charge on these four atoms are given

3.6 Conclusions

In this chapter, the Si-based micro-/nanointegrated fabrication technique presented in Chap. 2 is extended to be suitable for flexible materials, and a micro-/nanointegrated fabrication technique for flexible materials is developed, which is cost-efficient and universal. It contains two main steps, namely the controllable fabrication of ultra-low-surface-energy silicon mold with micro-/nanohierarchical structures and the single-step replication process for pattern transfer to the surfaces of flexible materials. In order to realize the above purpose, a fully optimized DRIE process is proposed by combining an improved DRIE process with a post-DRIE process, which is actually a single-step fabrication technique by optimizing the ICP process parameters. Based on this fully optimized DRIE process, the ultra-low-surface-energy silicon substrate with micro-/nanohierarchical structures can be controllably fabricated and used as the mold for single-step replication process of flexible materials.

Generally, a micro-/nanointegrated fabrication technique is proposed and successfully utilized to fabricate micro-/nanohierarchical structures on the surfaces of flexible materials. Two flexible materials including PDMS and parylene-C were micro-/nanostructured by this fabrication technique, and both of them were studied by designing two kinds of surface geometries, including pyramid shape and groove shape. The surface properties and the interaction of multiscale structures are explored by CA tests and FEA simulations, respectively. Additionally, the post-DRIE process is employed as a fluorocarbon plasma treatment process to enhance the hydrophobicity of flexible materials' surfaces, and the enhancement mechanism is deeply studied at the molecular level by using the DFT calculations for the first time.

In summary, based on the research work in both Chaps. 1 and 2, a universal micro-/nanointegrated fabrication technology is developed, which is cost-efficient, mass production, and suitable for silicon material as well as flexible materials. It can be used to fabricate wafer-level uniform micro-/nanohierarchical structures on silicon substrate or surfaces of flexible materials. The fabricated samples show several attractive properties, such as wide-band anti-reflectance, super-hydrophobicity, and SERS. This universal micro-/nanointegrated fabrication technology and the attractive properties from the fabricated samples open a new chapter of the development and applications of micro-/nanohierarchical structures.

References

1. Y. Yoon, D.W. Lee, J.H. Ahn, J. Sohn, J.B. Lee, in One-step fabrication of optically transparent polydimethylsiloxane artificial lotus leaf film using under-exposed under-baked photoresist mold. 25th *IEEE International Conference on Micro Electro Mechanical Systems*, January 29–February 2, Paris, France, 2012, pp. 301–304
2. H. Hassanin, K. Jiang, Multiple replication of thick PDMS micropatterns using surfactants as release agents. Microelectron. Eng. **88**, 3275–3277 (2011)
3. P. Campbell, M.A. Green, High performance light trapping textures for monocrystalline silicon solar cells. Sol. Energy Mater. Sol. Cells **65**, 369–375 (2001)
4. X.S. Zhang, Q.L. Di, F.Y. Zhu, H.X. Zhang, Wideband anti-reflective micro/nano dual-scale structures: fabrication and optical properties. Micro Nano Lett. **6**, 947–950 (2011)
5. P. Campbell, M.A. Green, Light trapping properties of pyramidally textured surfaces. J. Appl. Phys. **62**, 243–249 (1987)
6. Y. Suzuki, Y.C. Tai, Micromachined high-aspect-ratio parylene spring and its application to low-frequency accelerometers. J. Microelectromech. Syst. **15**, 1364–1370 (2006)
7. W. Li, D.C. Rodger, E. Meng, J.D. Weiland, M.S. Humayun, Y.C. Tai, Wafer-level parylene packaging with integrated RF electronics for wireless retinal prostheses. J. Microelectromech. Syst. **19**, 735–742 (2010)
8. E. Meng, P.Y. Li, Y.C. Tai, Plasma removal of parylene C. J. Micromech. Microeng. **18**, 145004 (2008)
9. B. Lu, D. Zhu, D. Hinton, M. Humayun, Y.C. Tai, Mesh-supported submicron parylene-C membranes for culturing retinal pigment epithelial cells. Biomed. Microdevices **14**, 659–667 (2012)
10. B. Lu, J.C.H. Lin, Z. Liu, Y.K. Lee, Y.C. Tai, in Highly flexible, transparent and patternable parylene-C superhydrophobic films with high and low adhesion. 24th *International Conference on Micro Electro Mechanical Systems* (*MEMS*2011), Cancun, Mexico, 2011, pp. 1143–1146
11. S. Wachsmann-Hogiu, T. Weeks, T. Huser, Chemical analysis in vivo and in vitro by Raman spectroscopy—from single cells to humans. Curr. Opin. Biotechnol. **20**, 63–73 (2009)
12. C. Krafft, G. Steiner, C. Beleites, R. Salzer, Disease recognition by infrared and Raman spectroscopy. J. Biophotonics **2**, 13–28 (2009)
13. H. Mao, W. Wu, D. She, G. Sun, P. Lv, J. Xu, Microfluidic surface-enhanced Raman scattering sensors based on nanopillar forests realized by an oxygen-plasma-stripping-of-photoresist technique. Small **10**, 127–134 (2014)
14. Y.S. Huh, A.J. Chung, D. Erickson, Surface enhanced Raman spectroscopy and its application to molecular and cellular analysis. Microfluid. Nanofluid. **6**, 285–297 (2009)
15. X.M. Lin, Y. Cui, Y.H. Xu, B. Ren, Z.Q. Tian, Surface-enhanced Raman spectroscopy: substrate-related issues. Anal. Bioanal. Chem. **394**, 1729–1745 (2009)

16. E.C.L. Ru, P.G. Etchegoin, Principles of Surface-Enhanced Raman Spectroscopy, (Elsevier B.V, Amsterdam, 2009)
17. J.P. Perdew, K. Burke, M. Ernzerhof, Generalized gradient approximation made simple. Phys. Rev. Lett. **77**, 3865–3868 (1996)
18. ADF 2010.01, http://www.scm.com
19. C.F. Guerra, J.G. Snijders, G. te Velde, E.J. Baerends, Towards an order-N DFT method. Theoret. Chem. Acc. **99**, 391–403 (1998)
20. G.T. Velde, F.M. Bickelhaupt, E.J. Baerends, C.F. Guerra, S.J.A. van Gisbergen, J.G. Snijders, T. Ziegler, Chemistry with ADF. J. Comput. Chem. **22**, 931–967 (2001)
21. E. van Lenthe, E.J. Baerends, Optimized slater-type basis sets for the elements. J. Comput. Chem. **24**, 1142–1156 (2003)
22. E. van Lenthe, E.J. Baerends, J.G. Snijders, Relativistic regular two-component Hamiltonians. J. Chem. Phys. **99**, 4597–4610 (1993)

Chapter 4
Flexible Triboelectric Nanogenerators: Principle and Fabrication

Abstract This chapter investigates the following five aspects: (I) It utilized the micro-nanointegrated fabrication technology mentioned above to fabricate nanogenerators and realized a novel high-performance sandwich-shaped triboelectric nanogenerator (TENG); (II) it established the theory model for three-layer TENG and obtained the numerical analysis of the device's electric output based on this model; (III) it thoroughly analyzed the working principle of sandwich-shaped TENG using the finite element method; (IV) it systematically tested its electric output performance and deeply studied the influence of frequency of applied force and the size of the device on device's output performance; and (V) it explored the performance of this sandwich-shaped TENG under different loads and validated its long-term stability and continuously working ability.

One of the effective ways to respond the global energy crisis and provide green sustainable energy is to harvest energy from the environment where we live. As a new energy conversion method, TENG with the advantages of high-output ability and pollution-free to the environment has attracted much attention, which has been mentioned in Sect. 1.3. However, most of the existed mechanical energy in our environment has very low frequency, which would make it very difficult to directly harvest and utilize this type of energy. Thereby, we introduced a novel sandwich-shaped and high-performance TENG in this chapter, which could be employed to convert the low-frequency mechanical energy into electric energy. In this design, an aluminum film is fixed between two PDMS film. This will allow sandwich-shaped TENG contact and separate twice under an applied pressure cycle, resulting in two electric output signals. The working principle of this sandwich-shaped TENG was simulated and studied by using the finite element method. To further enhance the output performance, the large-scale micro-/nanofabrication method was introduced to fabricate dual-scale structures to the PDMS surface. By the aid of the electric experiment platform, we systematically investigated the effects of

X.-S. Zhang, *Micro/Nano Integrated Fabrication Technology and Its Applications in Microenergy Harvesting*, Springer Theses, DOI 10.1007/978-3-662-48816-4_4

applied force's frequency and device's diameter on the generator's electric performance. Without any other electric circuits, this generator could directly illuminated 5 commercial light-emitting diodes (LEDs). This work moves the TENGs much closer to their practical applications.

4.1 Working Principle of TENG

In 1913, Neels Bohr proposed the famous Bohr model. This model depicts the atom as a small positively charged nucleus surrounded by electrons that move around the nucleus. Electrons and nucleus possess the same amount of charges but with opposite signs in atom. Therefore, the entire atom is electrically neutral. As is known to all, the matter is composed of orderly arranged atoms, so it also shows the property of electric neutrality. However, when two different materials are brought into contact with each other, the electrons would transfer from one material to the other one due to their distinct binding ability to the electrons. The effect that two different materials possess equal but opposite charges after touching is called triboelectrification, which is most significant to a triboelectric nanogenerator (TENG).

In 585 B.C., people observed this phenomenon. The philosopher Thales in the ancient Greek found that after silk and flannel rubbed with amber, they would possess the attractive ability to small objects, which is similar to lodestone. In our daily life, when rubber is touched against glass (or silk is touched against rubber), obviously contact-electrification phenomenon will be observed, as shown in Fig. 4.1. Although it was known three thousand years ago, how to effectively accumulate the produced charges in this process to realize a high-efficiency and

Fig. 4.1 The phenomenon of triboelectrification effect

minimized device and ultimately to supply power has puzzled the researchers for many years.

In January 2012, the research group led by Prof. Zhong Lin Wang from Georgia Institute of Technology firstly proposed a novel TENG [1]. They fabricated two thin metal foils as the electrode to accumulate charges at the back of two thin dielectric materials. Then, the two layers were placed together to form a TENG. Equal amount of charges with opposite signs (i.e., postive charges and negtive charges) will be generated by the contact of two different materials, and induced charges will be produced at the back of metal layer due to the electrostatic-induction effect in the separation process. As a result, the induced charge would flow from the outer circuit to another electrode and forms a current in the circuit.

Generally, the fundamental principle of TENG can be summarized as follows: Two different materials with back electrode are touched against each other, and equal but opposite charges will be generated. During the separating process, induced charges will occur on the electrode, and current will be produced at the outer circuit. Besides, the contact surface of material is also textured with some nanoscale structures. This will increase the surface roughness, thus enhancing the output performance. Therefore, it was called TENG. In addition, TENG can be divided into lateral sliding-mode TENG and vertical contact-mode TENG according to different contact ways [2].

The working principle of vertical contact-mode TENG is shown in Fig. 4.2. In the initial state, two friction layers of TENG are electrically neutral without charges on their surfaces. When the upper part is pressed to contact the lower part under a vertical force, equal but opposite charges will be generated at the contact surface, as shown in the Fig. 4.2(i). Then, upon the releasing of the force, the upper part will separate from the lower part, resulting in the increasing of the gap between two parts. Subsequently, an internal electric field will be established due to the separation of charges on the surfaces of two parts, and then opposite charges would be induced on the two electrodes (Fig. 4.2(ii)). Afterward, as the gap gradually increases, the induced charges on the two electrodes will increase in amount, and a maximum amount of charges will be obtained when it returns to the initial position, as shown in Fig. 4.2(iii). During this process, an electric potential difference will exist due to the opposite charges on the two electrodes. And a current will flow from one electrode to another as soon as two electrodes are connected via a load. When the vertical force is applied to the upper part again, the upper part will touch against the lower part repeatedly. This will cause the decreasing of charges on the two electrodes, and a reverse current will form in the external circuit loop, as depicted in Fig. 4.2(iv).

Figure 4.3 illustrates the working principle of sliding-mode TENG. When the upper part is pushed to rub the lower part under an in-plane sliding force, equal but opposite charges will be generated at the contact area, as shown in the Fig. 4.3(i). Then, the upper part will slide to the right under the sliding force toward right, which will cause the decrease of the overlap area of two materials. Due to the separation of charges on the surface of two parts, opposite charges are induced on the two electrodes (Fig. 4.3(ii)). Afterward, as the upper part gradually slides to the right, the induced charges on the two electrodes will be increasing.

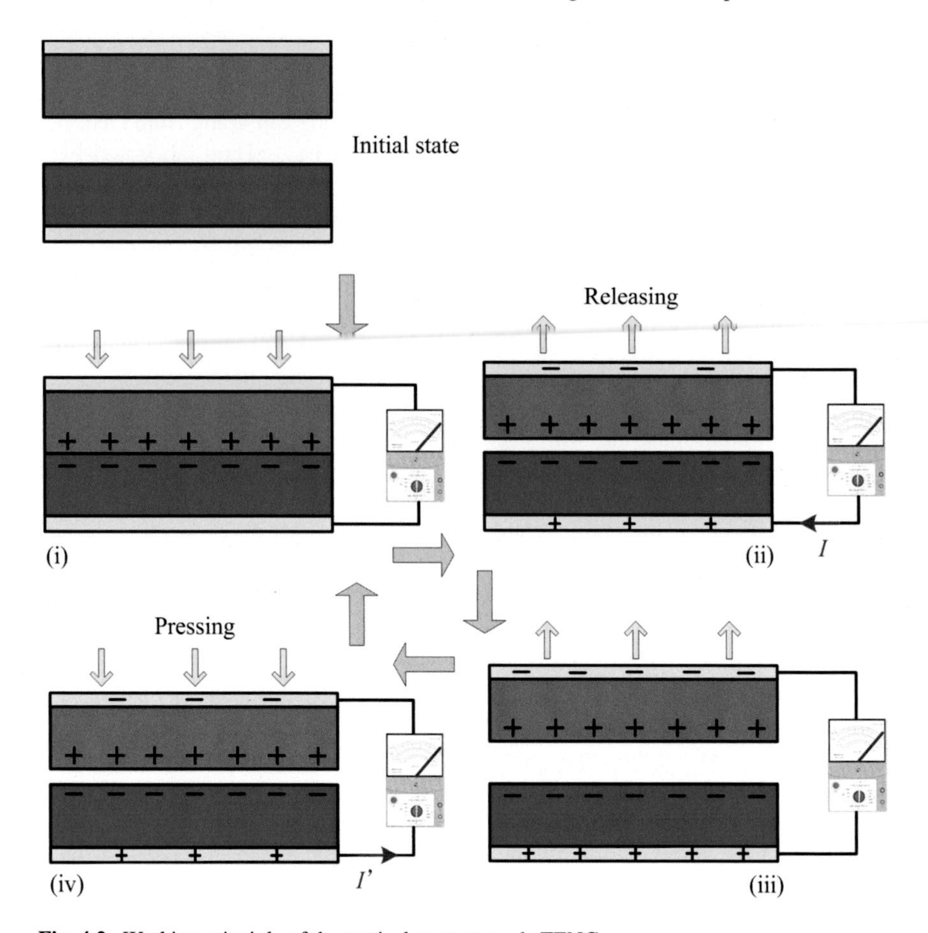

Fig. 4.2 Working principle of the vertical contact-mode TENG

And a maximum amount of charge will be obtained when the two parts are totally separated, as shown in Fig. 4.3(iii). During this process, an electric potential difference will exist due to the opposite charges on the two electrodes. And a current will flow from one electrode to another as soon as two electrodes connected via a load. When a reversed sliding force is applied to the upper part, the upper part will slide left against the lower part and finally return to its initial state. This will cause the decreasing of charges on the two electrodes, and a reverse current will form in the external circuit loop, as depicted in Fig. 4.3(iv).

The research group led by professor *Z.L. Wang* has presented a large amount of creative work in the field of TENG, who has developed several theoretical models for TENG based on different operating mechanisms. For the above-mentioned two types of TENG, they proposed a mathematical equation based on the output voltage V, the transferred charges Q, and the gap between two materials x, which is

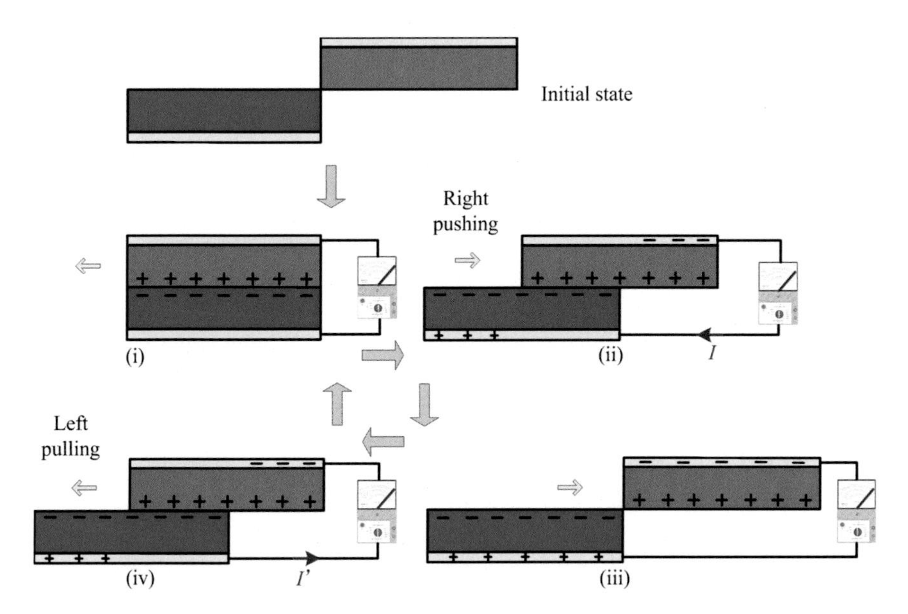

Fig. 4.3 Working principle of the sliding-mode TENG

also called as *V-Q-x* equation [3, 4]. The premise of this theoretical model is simplifying the real and complex device to a simple but reliable equivalent model.

In this model, the two contact materials can be regarded as two dielectric layers with different permittivity. Based on this assumption, TENG can be simplified by using the capacitor model. And the most important parameter during device's operation is the gap x. As a result, based on the physical model shown in Ref. [3], the *V-Q-x* theoretical equation for the vertical contact-mode TENG is given:

$$V = -\frac{Q}{S\varepsilon_0}\left(\frac{d_1}{\varepsilon_{r1}} + \frac{d_2}{\varepsilon_{r2}} + x(t)\right) + \frac{\sigma x(t)}{\varepsilon_0} \tag{4.1}$$

Meanwhile, Ref. [4] presented the *V-Q-x* theoretical equation for the lateral sliding-mode TENG:

$$V = -\frac{1}{\omega\varepsilon_0(l-x)}\left(\frac{d_1}{\varepsilon_{r1}} + \frac{d_2}{\varepsilon_{r2}}\right)Q + \frac{\sigma x}{\varepsilon_0(l-x)}\left(\frac{d_1}{\varepsilon_{r1}} + \frac{d_2}{\varepsilon_{r2}}\right) \tag{4.2}$$

Based on the physical equivalent mode, the derived theoretical equation is of significance to the development of the TENG, providing a theoretical support for further designing and optimizing high-performance TENGs.

4.2 Design of Flexible Sandwich-Shaped TENG

As discussed in Sect. 4.1, the basic geometry of TENG is a two-layer structure. It has two triboelectric materials and could produce one output signal, including a positive peak and a negative peak separately, under an external force in one cycle. So the frequency of produced electric signal is equivalent with that of the periodical external force. The advantage of this two-layer TENG is the simple structure and fabrication process, but its performance needs to be further enhanced. If the output from two devices could be added up by the aid of serial or parallel connection, its output performance can be easily increased. Thereby, we introduce a sandwich-shaped three-layer structure, which could produce two electric signals under external force in one cycle [5]. And its frequency would be twice of that of the applied forces. In other words, two outputs can be obtained from just one input. Besides, the well-designed micro-nano-dual-scale structures were employed to the contact surface of TENG. Compared with the traditional devices that use only microscale structures or nanoscale structures, the effective contact surface is further enlarged. Hence, the output performance can be significantly enhanced.

4.2.1 Structural Geometry and Surface Profile

Different from the existed two-layer TENG, the fabrication diagram of this novel sandwich-shaped TENG is shown in Fig. 4.4. The newly designed sandwich-shaped TENG includes three-layer triboelectric materials, which are the upper and lower thin PDMS films with micro-/nanostructures and the middle thin Al film. The entire device is fabricated to form an arch-shaped structure. The backside of the two PDMS layers is laminated with transparent and conductive PET/ITO and connected with metal lead to serve as an electrode of the TENG. The other electrode is extracted from the Al film in the middle. Moreover, the large-area micro-/nanointegrated fabrication technology is introduced to this novel TENG, realizing the PDMS surface with micro-/nanohierarchical structures. According to our previous analysis, increasing surface roughness would significantly increase the

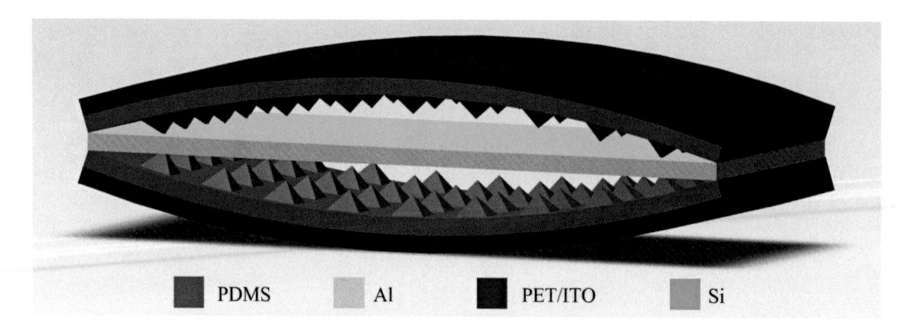

Fig. 4.4 The structural geometry of the sandwich-shaped TENG

effective contact area, resulting in a significant enhancement of the TENG's output performance. Therefore, micro-/nanohierarchical structures were introduced onto the surfaces of sandwich-shaped TENG to enhance the output performance. In addition, the structural design of sandwich shape realizes frequency multiplication by twice contact electrifications under one-cycle external force.

4.2.2 Theoretical Analysis

Traditionally, the TENGs contain two friction surfaces which can be employed to generate electric power by contact mode or sliding mode. In order to strengthen and functionalize TENG, the devices made of multiple friction surfaces were proposed [5–7]. Well-designed multiple structures can be utilized to realize multiple triboelectrification effects during one cycle of external force resulting in more power generation. However, the behavior of multiple-friction-surface TENG (i.e., MTEG) is more complicated, which is affected by extra factors, such as the internal interaction among multiple friction surfaces. Actually, this internal interaction may be ignored due to both of time difference of triboelectrification effects and continuous consumption of external load. Therefore, a simple but useful theoretical model for MTEG is established based on the *V-Q-x* relationship [3]. In order to simplify the explanation, three-layer contact-mode TENG is taken as a sample, as is shown in Fig. 4.5.

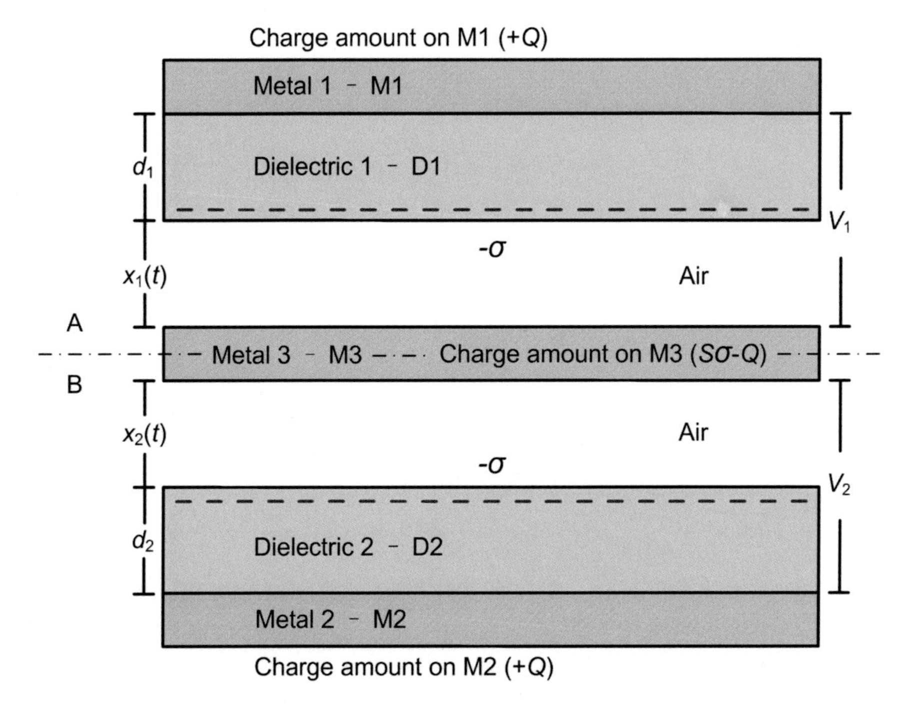

Fig. 4.5 The theoretical model of multilayer TENG

Obviously, the theoretical model shown in Fig. 4.5 is symmetrical, which can be divided into two same units, part A and part B. Herein, in order to simplify the calculation procedure, only one of the two symmetrical parts is analyzed. In part A, the thickness of dielectric layer 1 (D1) and its relative dielectric constant are set as d_1 and ε_{d1}, respectively. The area of two friction surfaces (i.e., D1 and M3) is set as S, and their distance is set as $x_1(t)$ that is variable under the external force. Therefore, when the external force is applied to the device, two frication surfaces will contact, and equal but opposite charges with the density of σ will be generated on the surfaces. Subsequently, when the external force is removed, two frication surfaces will separate, and a potential difference (V) will exist between them. Then, the transferred charges induced by the potential difference is set as Q, and in other words, the charge amount on metal 1 (M1) is Q. But for metal 3 (M3), it serves as not only the triboelectrification layer but also conductive electrode, so the charge amount on M3 consists of two parts, i.e., triboelectric charge ($S\sigma$) and transferred charge (Q).

Thus, the contribution of part A to the device output is actually defined by the potential difference (V_1) between M1 and M3. According to Fig. 4.5, V_1 is uniquely determined by the internal electric fields through D1 and air layer. Thus, the above internal electric fields are calculated as follows:

$$\text{Inside D1:}\, E_1 = -\frac{Q}{S\varepsilon_0\varepsilon_{d1}} \tag{4.3}$$

$$\text{Inside Air:}\, E_a = \frac{-\frac{Q}{S}+\sigma}{\varepsilon_0} \tag{4.4}$$

Thus, the potential difference V_1 is calculated by the combination of Eqs. 4.1 and 4.2:

$$V_1 = E_1 d_1 + E_a x_1(t) = -\frac{Q}{S\varepsilon_0}\left[\frac{d_1}{\varepsilon_{d1}} + x_1(t)\right] + \frac{\sigma x_1(t)}{\varepsilon_0} \tag{4.5}$$

According to the above theoretical analysis, similarly, the potential difference (V_2) of part B is calculated:

$$V_2 = -\frac{Q}{S\varepsilon_0}\left[\frac{d_2}{\varepsilon_{d2}} + x_2(t)\right] + \frac{\sigma x_2(t)}{\varepsilon_0} \tag{4.6}$$

Therefore, the potential difference for the whole three-layer (sandwich-shaped) TENG can be obtained by the combination of Eqs. 4.5 and 4.6:

$$V = V_1 + V_2 = -\frac{Q}{S\varepsilon_0}\left[\frac{d_1}{\varepsilon_{d1}} + \frac{d_2}{\varepsilon_{d2}} + x_1(t) + x_2(t)\right] + \frac{\sigma[x_1(t)+x_2(t)]}{\varepsilon_0} \tag{4.7}$$

The above theoretical Eq. 4.7 can also been extended to calculate n-layer MTEG by the following relation:

$$V = V_1 + V_2 + \cdots + V_n \tag{4.8}$$

It is worth mentioning that the above theoretical analysis is based on a simplified model without the internal interaction among multiple friction surfaces. Additionally, in practice, the time difference should be considered to obtain an accurate potential difference. However, this simplified theoretical model is still significant for predicting and improving the output performance of multilayer TENG.

4.2.3 Finite Element Simulation

In order to further study the operating mechanism of sandwich-shaped TENG, the finite element analysis (FEA) based on the software of COMSOL was used to simulate its working procedure, as shown in Fig. 4.6. At the origin state, no

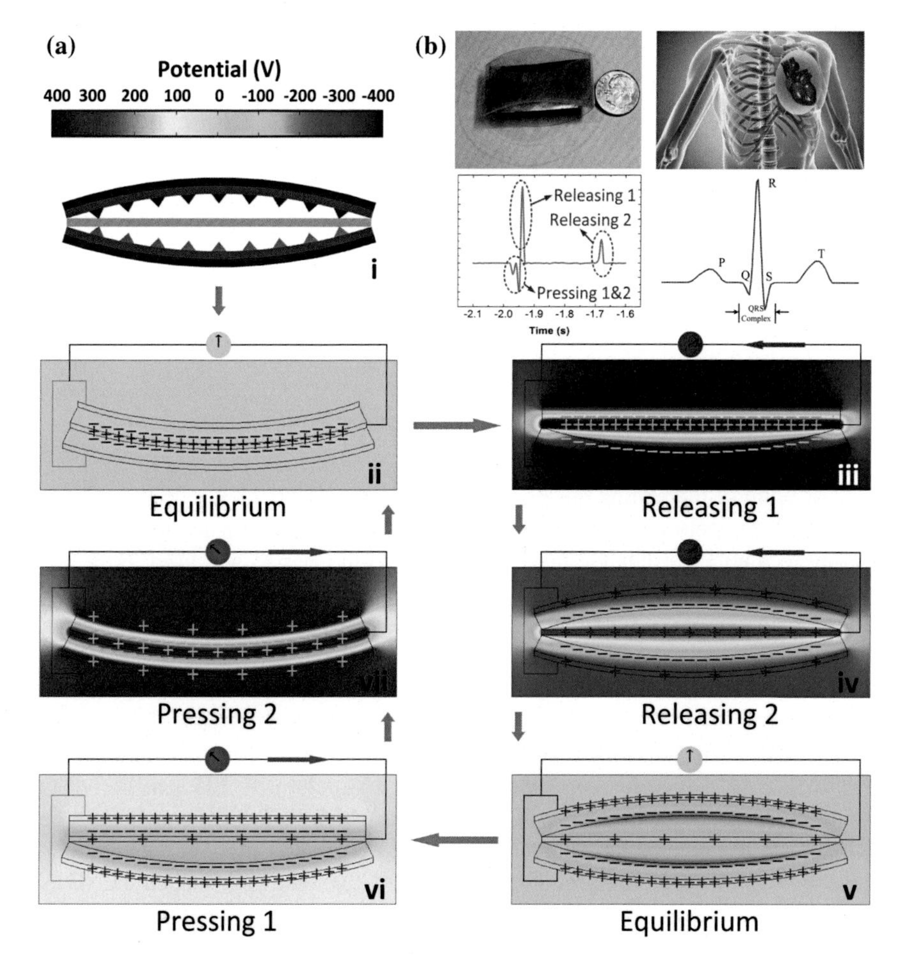

Fig. 4.6 **a** The FEA simulation of the working principle of the sandwich-shaped TENG, and **b** thc photograph of fabricated TENG and its electric signal in one cycle

electric potential exists between Al and PDMS as shown in Fig. 4.6a(i). By applying a periodic compressive force, Al and PDMS will rub twice during one cycle due to the specially designed sandwich-shaped structure, thus doubling the frequency of the output. When the device is pressed, the surfaces of Al and PDMS are charged with the same surface density shown in Fig. 4.6a(ii) [8]. As the force is removed, Al first separates from the bottom PDMS layer, and the potential difference between Al and PDMS will drive the electrons to flow through an external load (Fig. 4.6a(iii)). Soon afterward, the top PDMS layer separates from Al, generating a current flow with the same direction (Fig. 4.6a(iv)) until the device reaches electrical equilibrium (Fig. 4.6a(v)). When the generator is pressed again, the redistributed charges will build a reversed potential, thus driving electrons to flow in the opposite direction. Similarly, the friction between the top PDMS layer and Al gives the first peak output (Fig. 4.6a(vi)), and the rub of Al and bottom PDMS layer produces the second peak output (Fig. 4.6a(vii)).

It is worth mentioning that the downward peaks may be overlapped because of the rapid applied force. Additionally, the separating sequence of the layers fabricated by other materials could happen in other ways even for the same sandwich-shaped structure, which depends on the mechanical property [9]. Figure 4.6b shows the comparison between the output signal and electrocardiography (ECG) in one cycle. Two upward peaks with unequal amplitude can be clearly observed in the output signal, which is similar to the QRS complex and T wave in ECG [10]. By changing the gap between each layer, the interval between the upward peaks can be modulated to simulate the waveform of ECG, showing the potential application for the artificial component of heart.

4.3 Fabrication of the Sandwich-Shaped TENG

The key point of fabricating this novel sandwich-shaped TENG is realizing micro-/nanostructures onto the surfaces of PDMS. Thus, the micro-/nanointegrated fabrication technique for flexible materials presented in Chap. 3 was used. Here, the schematic view of this fabrication process flow is briefly summarized in Fig. 4.7, and more details can be found in Sect. 3.2.

Figure 4.8 shows the SEM images of the fabricated TENG, and the sandwich-shaped geometry is clearly shown in the cross-sectional view. An aluminum film with a thickness of 20 μm is fixed with elastic tape between two surface-micro-/nanostructured PDMS films with the thickness of 450 μm, which constitutes the effective contact-electrification sandwich-shaped structure. A 125 μm PET/ITO thin film was fabricated atop the PDMS film, in which PET layer was bent to be arch-shaped to enhance the output, and the ITO layer was used as the electrode to accumulate electric charges. The total size of the sandwich-shaped TENG is 2 cm × 4 cm.

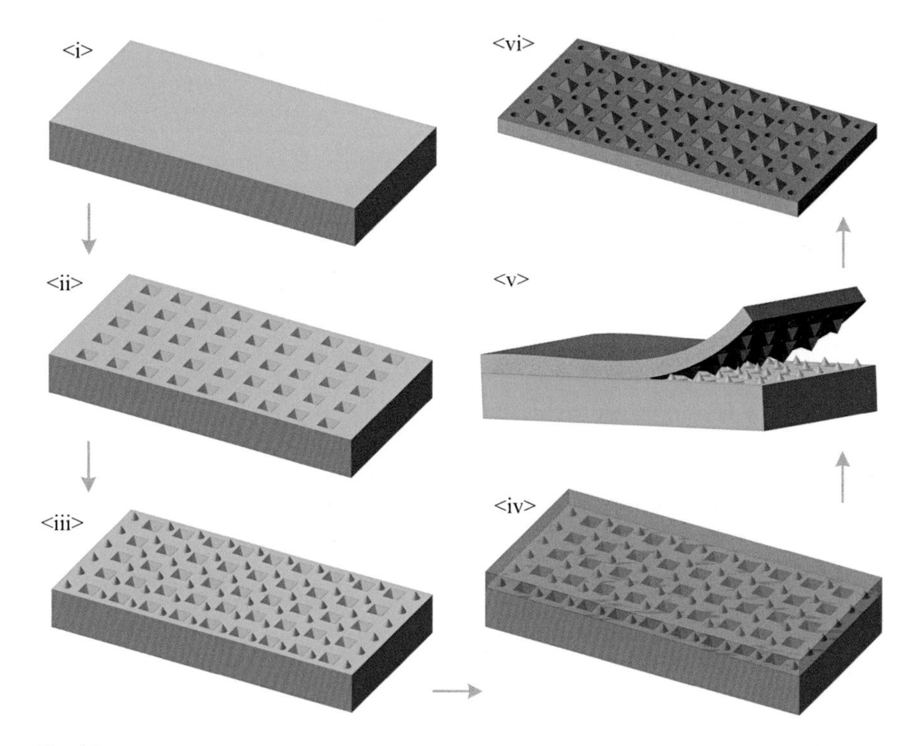

Fig. 4.7 The fabrication process flow of surface-textured PDMS films for the sandwich-shaped TENG

4.4 Electric Properties Test and Analysis of the Sandwich-Shaped TENG

4.4.1 Test System and Brief Results

In order to test the electric properties of the fabricated TENG, we set up a vibration platform that can be used to supply a controllable periodic external force with a specific frequency, as shown in Fig. 4.9. A sine wave with a designed amplitude and frequency was firstly generated by a waveform generator (RIGOL DG1022), and then, this sine signal was delivered to a power amplifier (SINOCERA YE5871A). Finally, the enlarged sine signal was supplied to drive and control the vibration of a shaker (SINOCERA). Thus, a cycled compressive force with controllable frequency was applied to impact the TENG. The output voltage and current of the sandwich-shaped TENG were measured via a digital oscilloscope (RIGOL DS1102E).

With pyramid-shaped micro-/nanohierarchical structures, the fabricated sandwich-shaped TENG achieved a maximum value, as shown in Fig. 4.10. The output

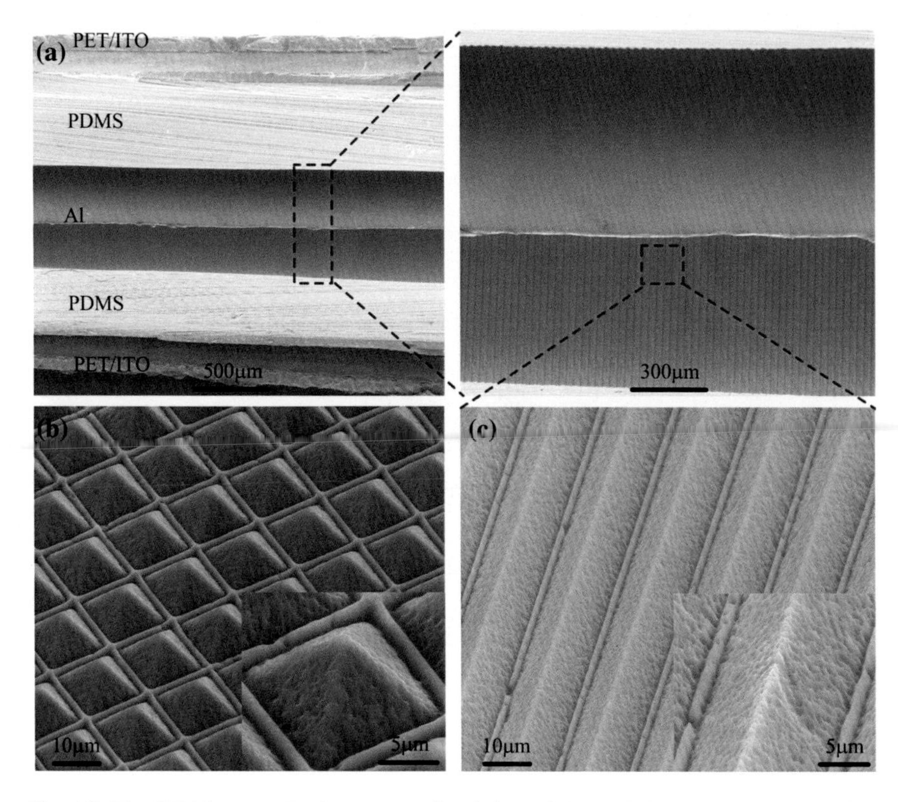

Fig. 4.8 The SEM images of **a** the cross-sectional view of the sandwich-shaped TENG and **b**, **c** the surface-micro-/nanostructured PDMS films

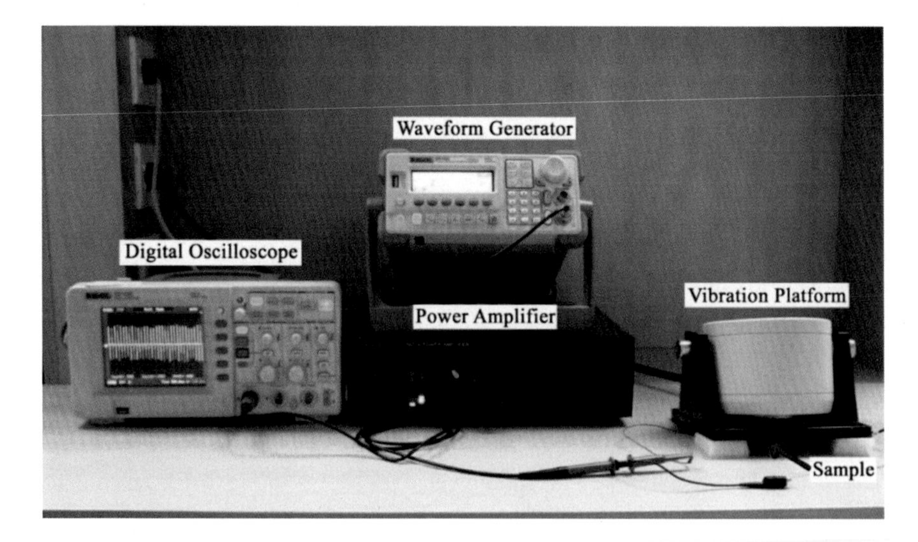

Fig. 4.9 The vibration platform for the electric test of the fabricated TENG

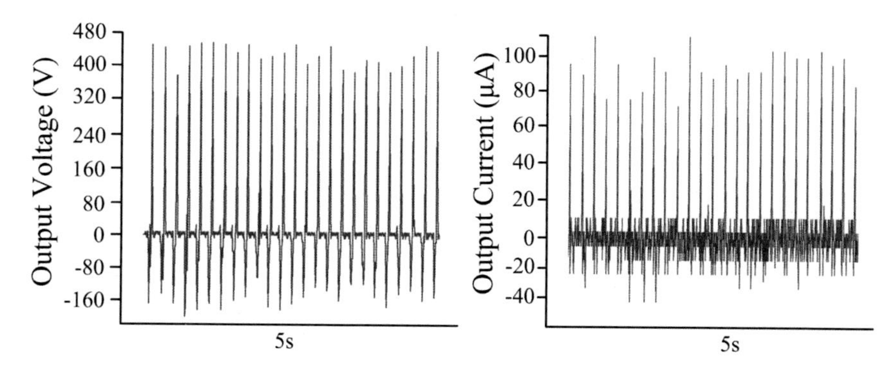

Fig. 4.10 The electric measurement of the fabricated TENG based on pyramid-shaped MNHS PDMS films

was measured by using an oscilloscope with a 100 MΩ probe, and the output current was measured by a low-value series resistance of 16 kΩ. The output voltage and current were measured to be 465 V and 107.5 μA, respectively. Thus, the current density and the power density achieved to 13.4 μA/cm^2 and 53.4 mW/cm^3, respectively.

4.4.2 Frequency Effect of External Force on TENG

The output performance of sandwich-shaped TENG clearly shows a frequency response to the external force. In other words, the frequency of external force affects the output performance of TENG, as shown in Fig. 4.11. Here, the sandwich-shaped TENG with surface-nanostructured PDMS was chosen to study its frequency response, and the frequency of external force increased from 1 to 10 Hz by a step of 1 Hz. When the frequency increased from 1 to 5 Hz, the output voltage of TENG continuously increased from 120 to 320 V, resulting from external electrons flow reaching equilibrium in a shorter time [8, 11]. When the frequency further increased from 5 to 7 Hz, the output voltage of TENG kept constant with an average value of 320 V. Subsequently, when the frequency increased from 7 to 10 Hz, the output voltage of TENG decreased to 218 V due to the under-releasing state of the sandwich-shaped TENG.

If the frequency of external force is too high, the cycle of external force is too short, which is defined as ΔT_f. The TENG cannot recover to the original position before the next force impact, when the cycle of external force is shorter than the recovery time of TENG, which is defined as Δt_r. This situation is defined as the under-releasing state of TENG, and the condition for this state is as follows:

$$\Delta T_f < \Delta t_r \tag{4.9}$$

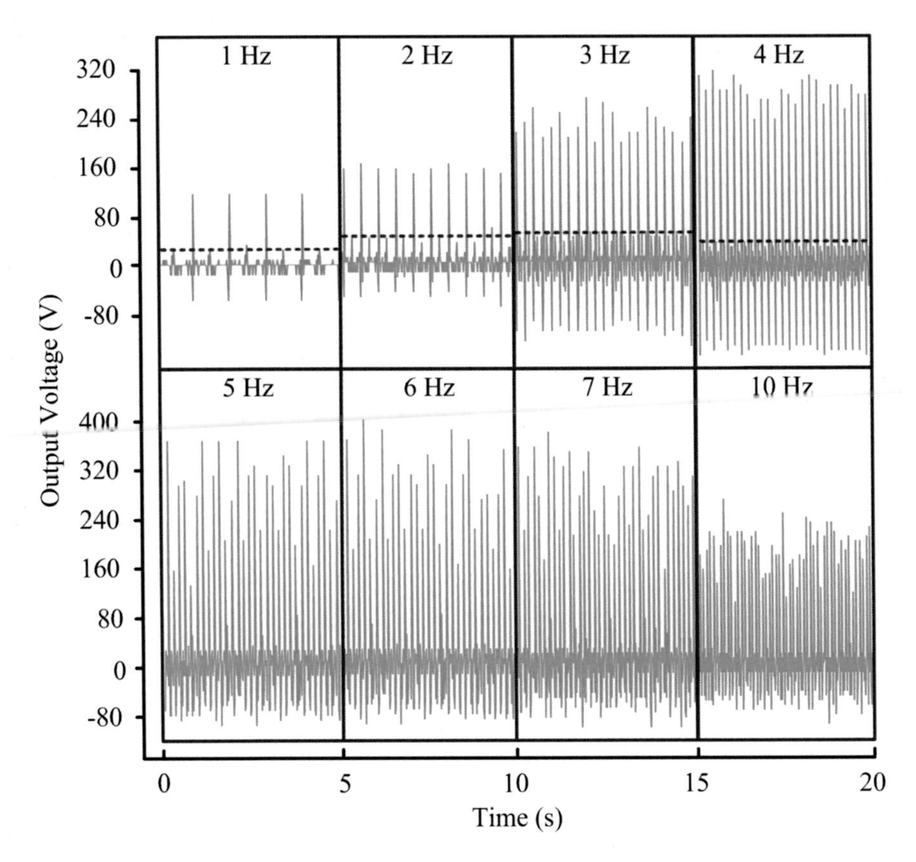

Fig. 4.11 Characterization of the output performance of the sandwich-shaped TENG with surface-nanostructured PDMS films under external forces with different frequencies in the range of 1–10 Hz

If the frequency of external force is defined as f, then the Eq. 4.9 is rewritten as follows:

$$f\left(=\frac{1}{\Delta T_f}\right) > \frac{1}{\Delta t_r} \tag{4.10}$$

Thus, a frequency threshold of f_0 is defined as the minimum frequency to make the TENG working in an under-releasing state. This mechanism also induces the trend of the frequency-multiplication output signal, whose peak output voltage is marked with red dashed line (Fig. 4.11). The maximum peak voltage of the frequency-multiplication output reached 62 V under 3 Hz external force.

4.4.3 Structural Effect on TENG

The structural geometry and parameters also affect the output performance of TENG, which can be summarized as follows.

(1) The power density of the generator is greatly influenced by the thickness of PDMS layer. As the charges in the electrode are induced, thinner PDMS layer will lead to more charges in the electrode, which means higher power density.
(2) The gap distance (planar separation distance) also influences the power density, which is already studied in Ref. [8]. The power density will increase with the gap distance but saturate when the gap distance increases to a certain value.
(3) The power density of the device is also influenced by the surface of the PDMS layer. According to our study, nanogenerator with micro-/nanodual-scale PDMS surface has the highest output performance due to the maximizing surface area of PDMS thin film.
(4) It has to be mentioned that the area of the device is not directly related to the power density. In the ideal case, the power increases linearly with the area, which means that the power density remains the same as the area changes. The charging capability has been tested by charging a 10 nF capacitor as shown in Fig. 4.12. The TENG made of flat PET films with the area of 2 cm × 2 cm has also been fabricated, whose charging capability is 10.6 nC

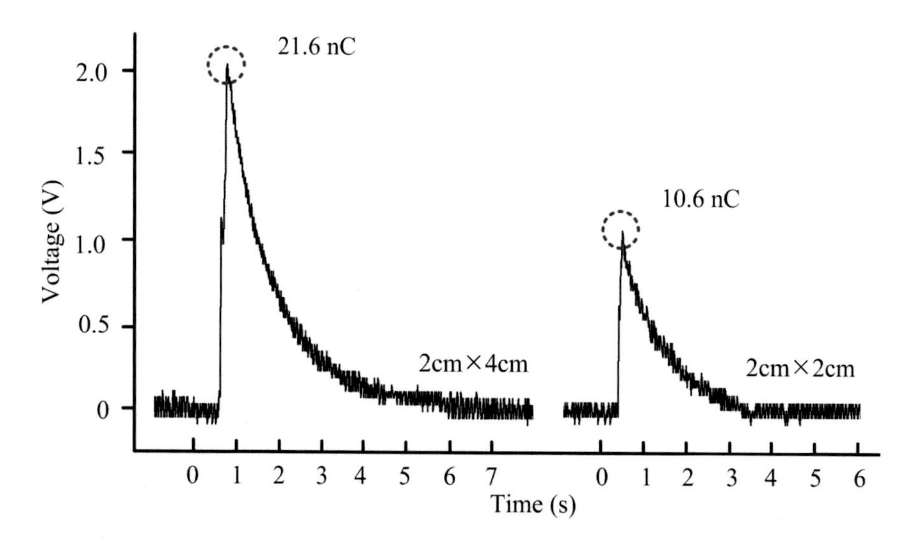

Fig. 4.12 The measured results of charging capability of TENG made of flat PET films, and the voltage is obtained from a 10 nF capacitor connecting to the TENG via a full-wave rectifier bridge

Fig. 4.13 The photographs of five paralleled LEDs directly powered by the sandwich-shaped TENG via hand pressing

in one cycle. While the charging capability of the double-area TENG (i.e., 2 cm × 4 cm) is 21.6 nC per cycle. Therefore, the ratio of the charging capability is 2.04, which almost equals to 2 (i.e., the ratio of the area).

4.4.4 Powering Ability for Practical Applications

According to the above electric property test, the fabricated sandwich-shaped TENG with micro-/nanohierarchical structures shows the remarkable output performance. Here, in order to verify the powering ability of this TENG for practical applications, it was demonstrated to lighten up 5 paralleled light-emitting diodes (LEDs) directly without any additional circuit, as shown in Fig. 4.13. When the TENG was compressed by hand, the instantaneous power was generated and lightened up five parallel-connected LEDs. Figure 4.13(i) shows that five LEDs are directly lightened without any energy storage unit or rectification circuit. Figure 4.13(ii) and (iii) shows the photographs of LEDs driven by the frequency-multiplication output (small peak) and the main output (large peak) during one external force cycle.

4.5 Conclusions

This chapter introduces the micro-/nanointegrated fabricated technology presented above into microenergy field, and a novel high-performance sandwich-shaped TENG was proposed. By fabricating micro-/nanohierarchical structures onto PDMS surface, the output performance of TENG was significantly enhanced. In addition, due to the sandwich geometry, the fabricated TENG generated two electric signals in one cycle of external force.

A theoretical model is established to analyze the multilayer TENG, while the FEA method is utilized to simulate the operating principle of sandwich-shaped TENG. Furthermore, the frequency effect of external force on TENG and the structural effect on TENG are studied by systematical electric tests. The output peak voltage, current density, and energy volume density achieve 465 V, 13.4 $\mu A/cm^2$, and 53.4 mW/cm^3, respectively. This sandwich-shaped TENG is demonstrated to be a robust micropower source by lightening up five paralleled LEDs directly.

References

1. F.R. Fan, Z.Q. Tian, Z.L. Wang, Flexible triboelectric generator. Nano Energy **1**, 328–334 (2012)
2. Z.L. Wang, Triboelectric nanogenerators as new energy technology for self-powered systems and as active mechanical and chemical sensors. ACS Nano **7**, 9533–9557 (2013)
3. S. Niu, S. Wang, L. Lin, Y. Liu, Y.S. Zhou, Y. Hu, Z.L. Wang, Theoretical study of contact-mode triboelectric nanogenerators as an effective power source. Energy Environ. Sci. **6**, 3576–3583 (2013)
4. S. Niu, Y. Liu, S. Wang, L. Lin, Y.S. Zhou, Y. Hu, Z.L. Wang, Theory of sliding-mode triboelectric nanogenerators. Adv. Mater. **25**, 6184–6193 (2013)
5. X.S. Zhang, M.D. Han, R.X. Wang, F.Y. Zhu, Z.H. Li, W. Wang, H.X. Zhang, Frequency-multiplication high-output triboelectric nanogenerator for sustainably powering biomedical microsystems. Nano Lett. **13**, 1168–1172 (2013)
6. B. Meng, W. Tang, X. Zhang, M. Han, W. Liu, H. Zhang, Self-powered flexible printed circuit board with integrated triboelectric generator. Nano Energy **2**, 1101–1106 (2013)
7. G. Zhu, P. Bai, J. Chen, Z.L. Wang, Power-generating shoe insole based on triboelectric nanogenerators for self-powered consumer electronics. Nano Energy **2**, 688–692 (2013)
8. S. Wang, L. Lin, Z.L. Wang, Nanoscale triboelectric-effect-enabled energy conversion for sustainably powering portable electronics. Nano Lett. **12**, 6339–6346 (2012)
9. G.M. Rebeiz, *RF MEMS: theory, design, and technology* (Wiley, New York, 2003), pp. 21–86
10. J.W. Hurst, Naming of the waves in the ECG, with a brief account of their genesis. Circulation **98**, 1937–1942 (1998)
11. L. Gu, N.Y. Cui, L. Cheng, Q. Xu, S. Bai, M.M. Yuan, W.W. Wu, J.M. Liu, Y. Zhao, F. Ma, Y. Qin, Z.L. Wang, Flexible fiber nanogenerator with 209 V output voltage directly powers a light-emitting diode. Nano Lett. **13**, 91–94 (2013)

Chapter 5
Flexible Triboelectric Nanogenerators: Enhancement and Applications

Abstract In this chapter, the density functional theory is employed to analyze the triboelectrification effect, which reveals the underlying mechanism of the materials' abilities of capturing or losing electrons at the molecular level for the first time. Three universal approaches, including single-step plasma treatment, surface texturing-based micro-/nanohierarchical structures, and hybrid mechanisms, are proposed to enhance the output performance of triboelectric nanogenerators. In addition, the applications of triboelectric nanogenerators for self-powered sensors, commercial electronics, and biomedical microsystems are successfully demonstrated.

As mentioned in Sect. 1.3.3, the output performance of triboelectric nanogenerator (TENG) has been gradually improved during the past two years, which can be used to power electronic microdevices and systems as well as low-power lamp. However, there are still great challenges for the rapid development of TENGs, and the most important two key points are fundamental principle exploration and output power density enhancement. Although the literal records of triboelectricity appeared 3000 years ago, the underlying mechanism of triboelectrification effect is still not clear. Besides, although output power of TENGs can be enhanced by enlarging size and optimizing structure, the power density still needs to be improved to meet the energy requirements of electronic devices and systems in practical applications.

5.1 Enhancement of TENG Based on Single-Step Fluorocarbon Plasma Treatment

Here, we introduce the micro-/nanointegrated fabrication technology presented above into the TENG fabrication process [1]. The output performance of TENG was greatly improved with the output power density enhanced 2.78 times by depositing fluorocarbon polymer on the surface of friction materials. This method on the

X.-S. Zhang, *Micro/Nano Integrated Fabrication Technology and Its Applications in Microenergy Harvesting*, Springer Theses, DOI 10.1007/978-3-662-48816-4_5

basis of single-step fluorocarbon plasma treatment has the following advantages: (1) simple and effective; (2) universal and applicable for many friction materials, such as Kapton, PET, PDMS, and Al; (3) single-step process with low cost; (4) applicable for many types of surface structures including micro-/nanohierarchical structures; and (5) good large-area controllability and compatibility.

5.1.1 Structural Design and Fabrication

The rapid separation between triboelectric pairs is very important for the output performance of TENG, and therefore, the arch-shaped structure is widely used in the design of high-performance TENGs [2]. Herein, this TENG was also designed to be arch-shaped by using two bended aluminum (Al) foils as the supporting structure and electrodes, as shown in Fig. 5.1a. This upper Al foil also serves as one of the triboelectric pair. On the lower Al foil, a PDMS film with micro-/nanohierarchical structures was placed as the other of triboelectric pair, which was fabricated by the micro-/nanointegrated fabrication technology proposed above in this thesis, as shown in Fig. 5.1b. Figure 5.2 shows the photograph of the fabricated TENG and SEM images of PDMS with pyramid-shaped micro-/nanohierarchical structures.

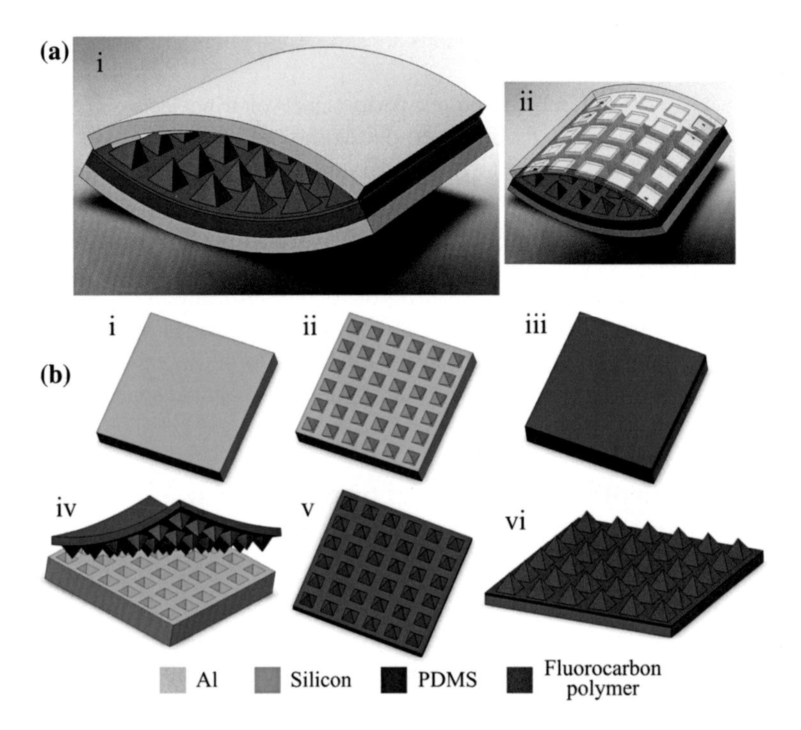

Fig. 5.1 Schematic view of **a** 3-D structure and **b** fabrication process flowchart of the high-performance triboelectric nanogenerator

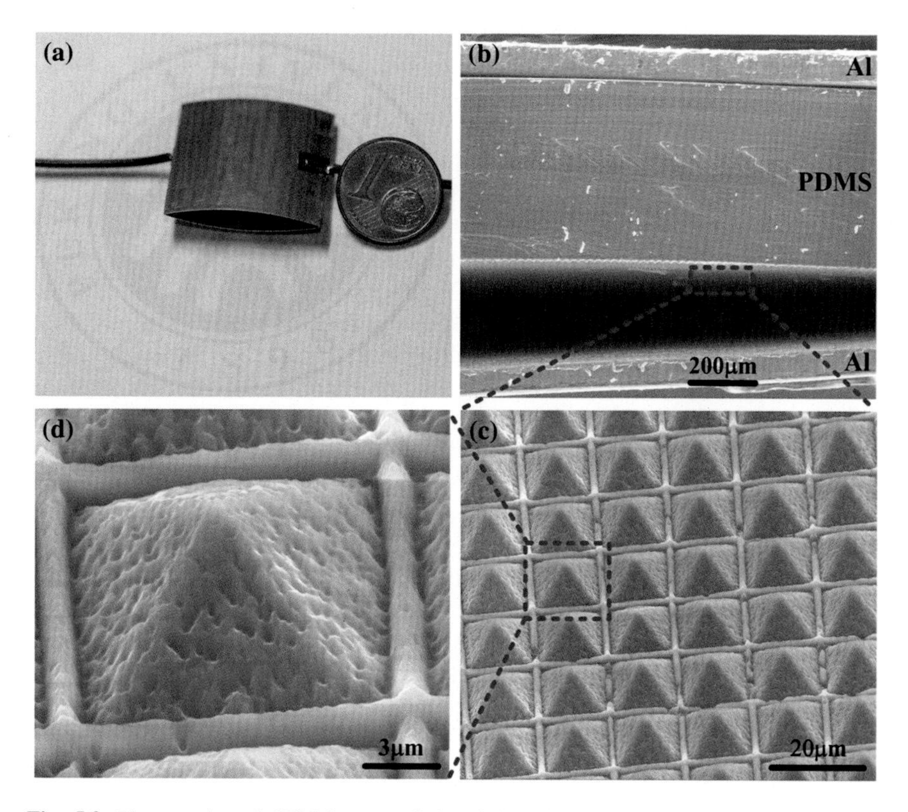

Fig. 5.2 Photograph and SEM images of the high-performance triboelectric nanogenerator. **a** Photograph and **b** cross-sectional view of the device; **c**, **d** SEM images of PDMS surface with pyramid-shaped micro-/nanohierarchical structures

5.1.2 Triboelectric Mechanism Analysis Based on DFT Calculations

Section 5.4.1 summarizes the development of theoretical analysis of triboelectric nanogenerator, and most of them focus on establishing the equivalent circuit model. Previously, the theoretical analysis of triboelectric nanogenerator was built up basically at the device level. However, the core mechanism of TENG, i.e., triboelectrification effect, is still not clear. As shown in Table 1.1, the greater the relative difference between two materials in the triboelectric series, the more charges the triboelectrification effect will produce. However, the underlying mechanism that determines the relative position of materials in the triboelectric series still needs to be further explored. In this section, we introduced the density functional theory (DFT) into the mechanism analysis of triboelectrification effect and performed theoretical calculations to evaluate the ability of capturing or losing electrons of common triboelectric pairs at the molecular level.

According to previous work [1, 3–5], the voltage and current outputs of TENG tightly correlate with the amount of transferred charges (q), where

$$q = S \times \Delta\sigma \tag{5.1}$$

and S is the surface area of triboelectric layer, and $\Delta\sigma$ is the transferred charge density. Basically, q is determined by the triboelectric charge density (σ_0) and the separation distance (d) of triboelectric pair. The maximum separation distance d_0 is determined by the device geometry and the material property, which can be considered as a constant for a specific TENG. The triboelectric charge density σ_0 is related to the relative difference of triboelectric pair in the triboelectric series. In other words, σ_0 depends on the relative difference of capturing electrons, which can be evaluated by the electron binding energy. Therefore, the semiquantitative analysis on the ability of capturing or losing electrons of triboelectric pair can be carried out by investigating their electron detachment energies.

We study the relative ability of losing one electron between fluorocarbon polymer deposited by C_4F_8 plasma and PDMS by calculating the vertical ionization energies of two model complexes $CF_3(CF_2–CF_2)_2CF_3$ and $(CH_3)_3Si(OSi(CH_3)_2)_2OSi(CH_3)_3$, respectively, at the DFT level. We employed the generalized gradient approximation (GGA) with the PBE exchange–correlation functional and the hybrid functional B3LYP as implemented in the Amsterdam Density Functional (ADF 2010.02) program. The Slater basis sets with the quality of double-ζ plus one polarization functions (DZP) [6–11] were used, with the frozen core approximation applied to the inner shells $[1s^2 - 2p^6]$ for Si and $[1s^2]$ for C, O, and F. The scalar relativistic (SR) effects were taken into account by the zero-order-regular approximation (ZORA) [10]. Geometries of neutral $CF_3(CF_2–CF_2)_2CF_3$ and $(CH_3)_3Si(OSi(CH_3)_2)_2OSi(CH_3)_3$ were fully optimized, and vibrational frequencies were calculated to verify the local minima on the energy surface at the SR-DFT/PBE level. To calculate the vertical detachment energies of the two neutral molecules above, the B3LYP single-point energy calculations were performed on the corresponding neutral and cationic molecules at the DFT/PBE ground-state geometries of neutrals.

For the convenience of the statement, we label the two model complexes as $M_{C_4F_8}$ and M_{PDMS}, respectively. Their optimal structures and vertical ionization energies are shown in Fig. 5.3. Clearly, complex $M_{C_4F_8}$ has much larger ionization energy than complex M_{PDMS}, i.e., 12.31 eV versus 8.98 eV. This clearly shows that the former is more difficult to lose one electron than the latter, consistent with the observed experimental results of enhanced TENG output performance by C_4F_8 plasma treatment.

To validate the reliability of the above theoretical methods, we measured the Fourier transform infrared (FTIR) spectra of the fluorocarbon layer and calculated the corresponding FTIR spectra of the above model complexes. As shown in Fig. 5.4, the calculated FTIR spectra are highly consistent with the measured one, confirming the reliability of the above theoretical calculations. The strong absorption band near 1200 cm^{-1} and the weak absorption band near 700 cm^{-1} indicate the coupling stretching vibration of C–F single bond and C–C single bond [12, 13].

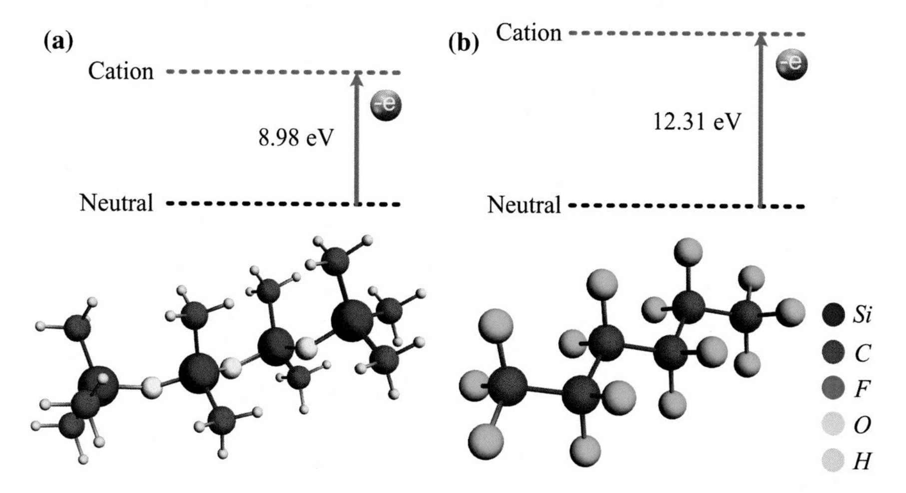

Fig. 5.3 The theoretical calculation results of vertical ionization energy of model complexes of **a** PDMS and **b** fluorocarbon layer deposited by the C_4F_8 plasma treatment

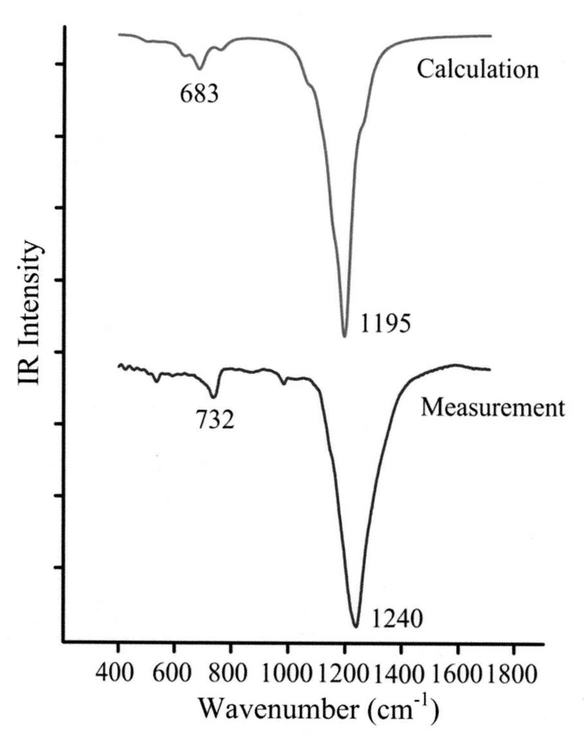

Fig. 5.4 Calculated and measured Fourier transform infrared (*FTIR*) spectra of the fluorocarbon layer deposited by post-DRIE process with the passivation gas of C_4F_8

As mentioned above, This single-step fluorocarbon plasma treatment process based on DRIE process enhances the output performance of TENG and realized the multiplication of output power density.

5.1.3 Effect of Fluorocarbon Plasma Treatment on Output Performance of TENG

According to the measurement results shown in Fig. 5.5, the output performance of TENG is significantly enhanced by the fluorocarbon plasma treatment. It is confirmed that the reaction time (i.e., the cycle of plasma treatment) is the most important factor affecting the fluorocarbon plasma treatment based on a number of comparative experiments. Comparing sample (a) with (b) in Fig. 5.5(i), (ii), it clearly shows that the output peak voltage of TENG treated for one cycle (i.e., 12 s) increases sharply from 124 to 193 V.

The output voltage of TENG increases continuously as the cycle of plasma treatment increases up till 8 cycles. The output peak voltage reaches to a maximum value of 265 V after 8-cycle plasma treatment, which is improved by a factor of 214 % compared to that of untreated sample (a). The instantaneous output current is also enlarged up to 73 μA by a 8-cycle plasma treatment voltage, an

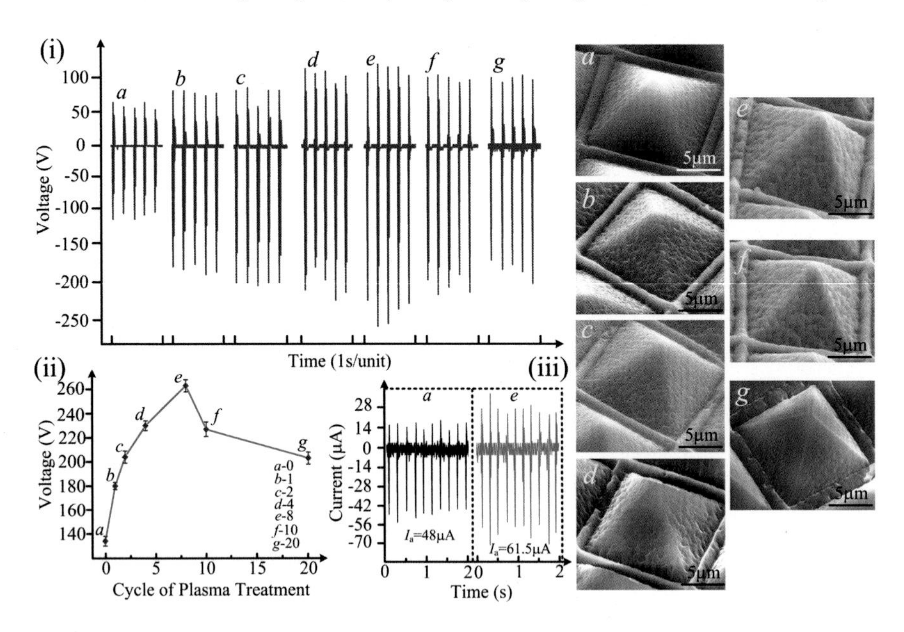

Fig. 5.5 Characterization of the output performance of TENG treated with different plasma treatment recipes (i.e., plasma treatment cycle) under a 5 Hz external force. (**i**) Output voltage waveform; (**ii**) Curve of output voltage versus plasma treatment cycle; (**iii**) Output current waveform of sample *a* and *e*. (*a*-0 cycle, *b*-1 cycle, *c*-2 cycles, *d*-4 cycles, *e*-8 cycles, *f*-10 cycles, *g*-20 cycles; one cycle lasts 12 s)

increase by a factor of 130 % in comparison with 56 μA of the sample (a), as shown in Fig. 5.5(iii). During the plasma treatment, the area of the sample is kept constant. And therefore, the energy volume density of TENG is enhanced by a factor of 278 % by using the optimized parameter of 8 cycles (i.e., 96 s).

It is worth mentioning that the fluorocarbon plasma treatment brings not only the chemical modification to enhance the TENG output performance but also the physical effect on the morphology of the surface-textured PDMS film, as shown in SEM images in Fig. 5.5a–g. The surface-textured PDMS film was gradually covered by a thin fluorocarbon layer as the number of the plasma treatment cycle increases from one to eight (i.e., the time of the plasma treatment increases). In other words, the PDMS film with micro-/nanohierarchical structures was partly covered by the fluorocarbon layer in the cycle ranging from 1 to 7, while were completely covered after 8-cycle plasma treatment. Subsequently, the thickness of the fluorocarbon layer deposited on the surface-textured PDMS is gradually increasing. While according to the analysis in Sect. 4.1, the thicker dielectric layer reduces electric output of the TENG due to the weakened electrostatic induction. But considering that the thickness of PDMS is 10^2–10^3 order larger than that of fluorocarbon polymeric layer, the negative effect of fluorocarbon layer thickness increase on the TENG output performance is very weak.

In addition, by comparing the SEM images among samples *a*–*g* in Fig. 5.5, the surface roughness of PDMS film decreases as the cycles of plasma treatment increase. After 20-cycle plasma treatment, the sample's surface becomes smooth and the nanostructures are completely covered by the fluorocarbon polymer. According to previous work in the literature, the TENG output significantly decreases as the surface roughness decreases.

As mentioned above, the fluorocarbon plasma treatment brings two effects, chemical enhancement and physical diminution, on the TENG output performance. And therefore, there exists a balance point under these two effects, i.e., the optimal value of TEMG output performance. According to Fig. 5.5(ii), the TENG output voltage enlarges as the cycle of plasma treatment increases from 0 to 8 (i.e., point *a*–*e*), and the change rate (i.e., the curve's slope) gradually decreases. An opposite change trend is observed when the cycle number is large than 8 (i.e., point *e*–*g*). Thus, the optimized parameters of plasma treatment (i.e., eight cycles) are obtained.

5.1.4 Reliability of Fluorocarbon Plasma Treatment for Enhancing TENG

We further investigated the reliability and generality of the fluorocarbon plasma treatment from three aspects, as shown in Figs. 5.6 and 5.7.

Firstly, the contribution of fluorocarbon polymer to the TENG performance enhancement is confirmed by comparison with the noble gas (i.e., Ar) plasma treatment, as is shown in zone *A* of Fig. 5.6. The output voltages before and after

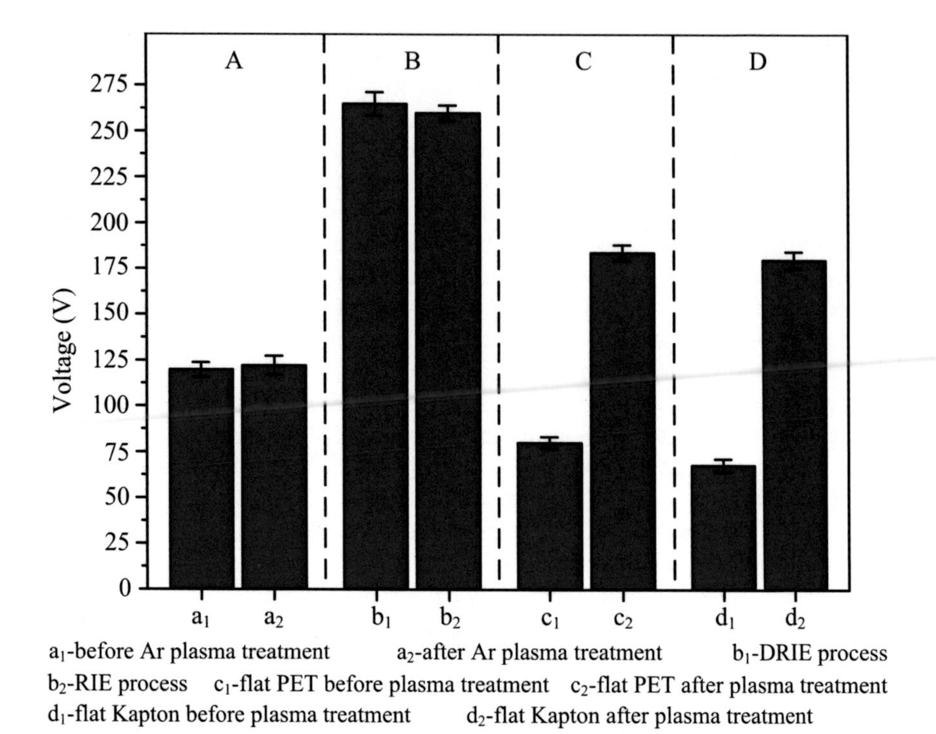

Fig. 5.6 Investigation of the reliability of the fluorocarbon plasma treatment including (**A**) different plasma molecules, (**B**) different plasma treatment approaches, (**C**, **D**) different substrate materials

Ar plasma treatment almost do not change, i.e., 122 V versus 124 V, showing that the plasma itself does not affect the PDMS surface. And therefore, the fluorocarbon polymeric material deposited by the C_4F_8 plasma treatment is the core of enhancing TENG performance by fluorocarbon plasma treatment.

Secondly, we employed a different fluorocarbon plasma treatment method, the reactive-ion etching (RIE) process, to demonstrate the reliability of fluorocarbon plasma treatment for improving the TENG performance. The difference between RIE process and DRIE process is illustrated in Fig. 2.1 in Chap. 2. By using the same C_4F_8 gas, the output voltages of the RIE-treated TENG and the DRIE-treated TENG achieved to 260 and 265 V, respectively, as shown in zone *B* of Fig. 5.6. Clearly, this shows that different plasma treatment processes with the same gas achieve almost the same enhancement level, confirming the reliability of fluorocarbon plasma treatment.

Finally, this fluorocarbon plasma treatment is also applied to another two common substrate materials, PET and Kapton, to further demonstrate the enhancement ability of this method. After the fluorocarbon plasma treatment, the output voltage of TENG increased from 80 to 184 V for the flat PET substrate and from 60 to 184 V for the flat Kapton substrate, as is shown in *Zone C* and *D* in Fig. 5.6.

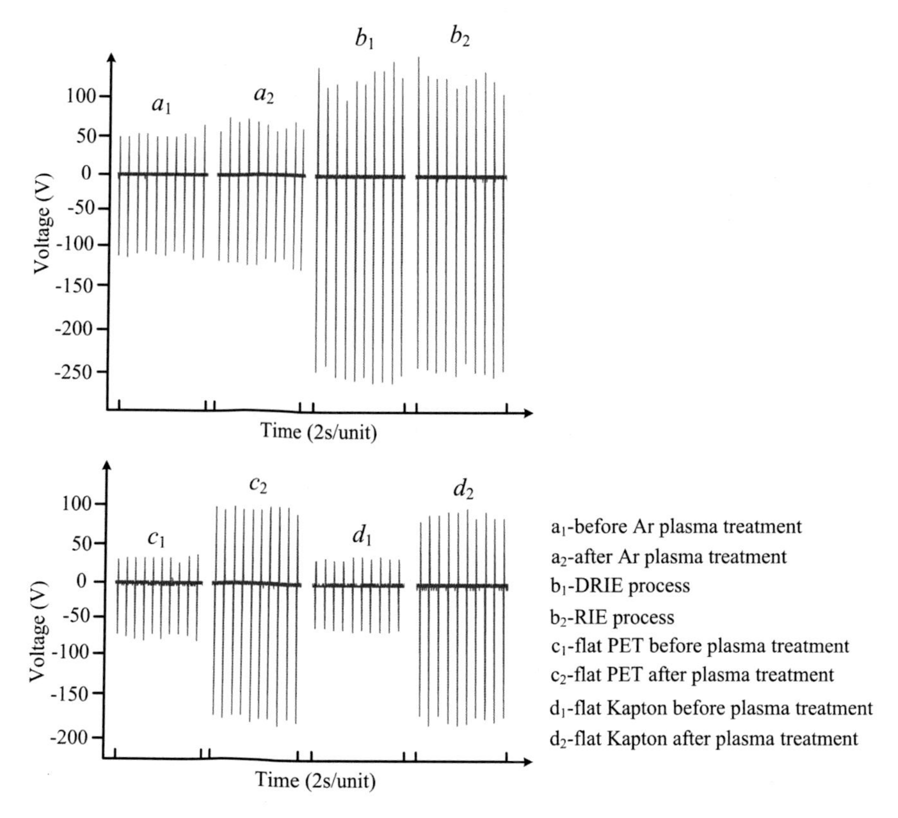

Fig. 5.7 The detailed output voltage waveforms in the reliability test of fluorocarbon plasma treatment shown in Fig. 5.6

Based on the above systematic investigation, this fluorocarbon plasma treatment is demonstrated to be a reliable and universal approach to enhance the TENG output performance.

5.1.5 The Performance Enhancement Analysis

This subsection is to analyze the basic electric characteristics and stability of fluorocarbon plasma-treated TENG and the environmental effect on the output performance. Figure 5.8 shows the measured voltage waveform of TENG during one pressing–releasing cycle under the external force by using an oscilloscope with a 100 MΩ probe. Although the resistance value of the probe is as high as 100 MΩ, the voltage measurement circuit cannot be considered as an open-circuit one due to the high-value internal resistance of TENG, which is usually at tens MΩ level. According to the area integral of the curve in the Fig. 5.8, the transferred charges from the negative and positive electrodes have the same amount with opposite polarity in one cycle.

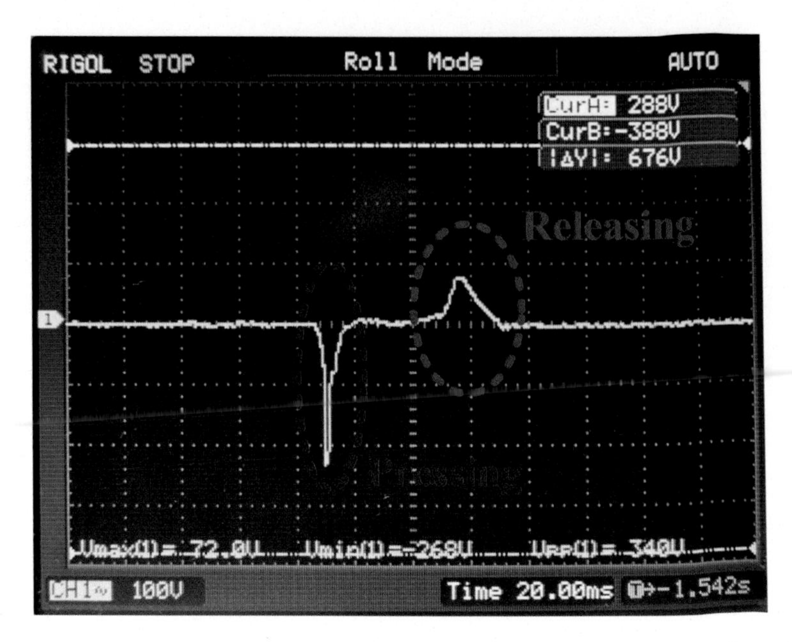

Fig. 5.8 The output voltage waveform of TENG during one pressing–releasing cycle under an external force

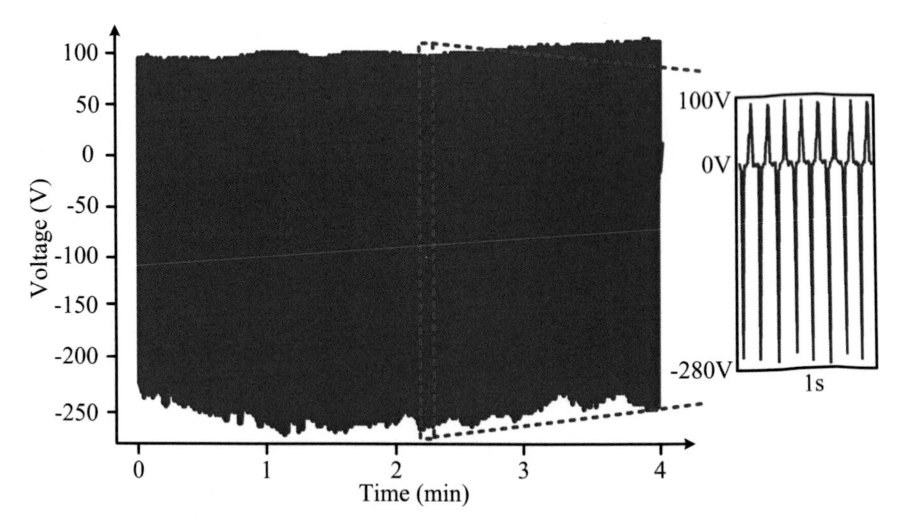

Fig. 5.9 Stability test of the plasma-treated TENG in about 2000 continuous cycles of 8 Hz external force

That is to say, the charges produced by the Al and PDMS friction pair under the triboelectrification effect have opposite polarities and the same amount.

We investigate the stability of TENG performance after fluorocarbon plasma treatment by measuring the output voltage waveform under a continuous working

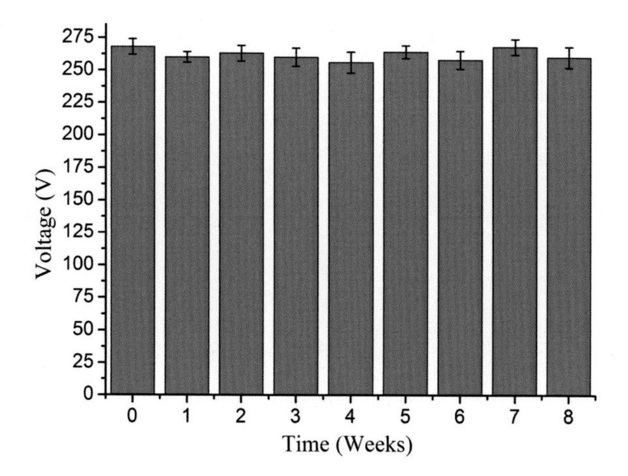

Fig. 5.10 Stability test of the change of the plasma-treated TENG output performance as the time increases after the fluorocarbon plasma treatment

state and the output voltage in a long period, as shown in Figs. 5.9 and 5.10, respectively. By using the measurement system illustrated in Fig. 4.9, the reliability of the fabricated TENG was tested by 2000-cycle continuous external force (8 Hz for 4 min). Clearly, the plasma-treated TENG works very well continuously for about 2000 cycles, and the output voltage peak keeps at 250 V. The measurement results in Fig. 5.10 show that the output voltage is very stable and almost constant during the observation time of 8 weeks. The voltage difference is ±6 V, which can be actually ignored considering the measurement error of a few volts (see error bar in Fig. 5.10).

5.2 Enhancement of TENG Based on Hierarchical Structures

5.2.1 Enhancement of TENG Based on Micro-/Nanohierarchical Structures

To investigate the influence of triboelectric surface profile on the TENG performance, we designed three types of surface structures, including nanostructures, microscale periodic arrays, and micro-/nanohierarchical structures, where the latter two ones contain both of pyramids and V-shaped grooves. The SEM images of the fabricated PDMS films are shown in Fig. 5.11a–g, and their silicon molds are shown in Fig. 5.12.

Figure 5.11 compares the electric output performance of sandwich-shaped TENGs with different surface profiles. In general, the output voltage and the current of TENG are enhanced by increasing the surface roughness. In this thesis, micro-/nanohierarchical structures (Fig. 5.11e, g) are firstly introduced into TENG, which further strengthen the TENG performance compared with pure

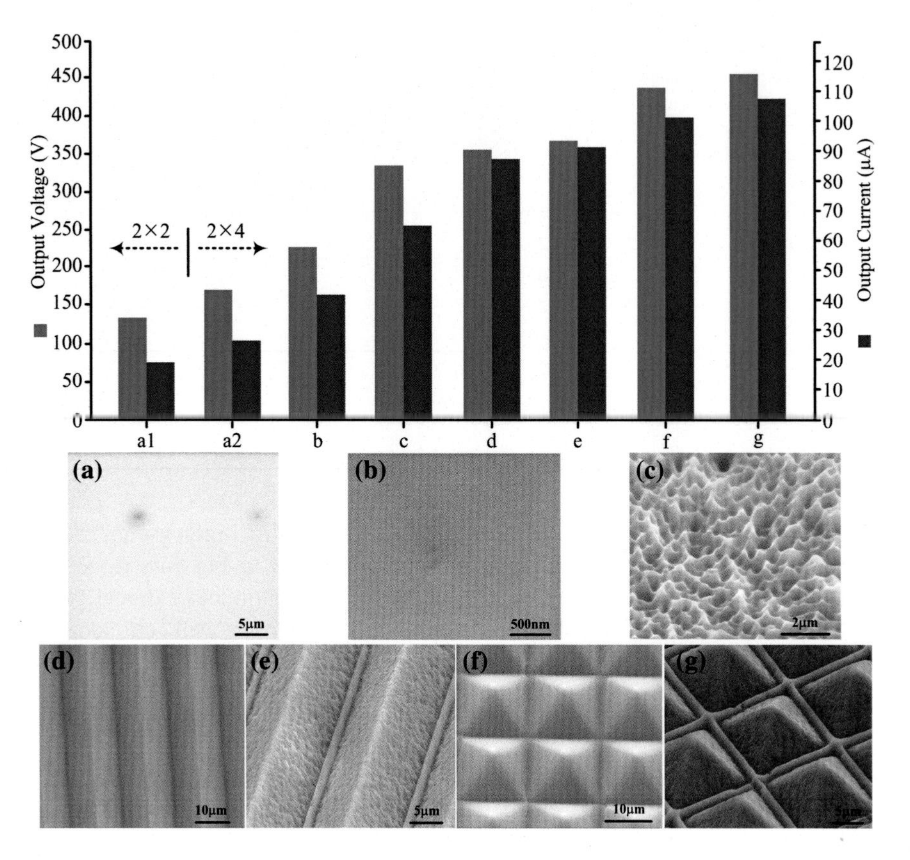

Fig. 5.11 Characterization of the output performance of the sandwich-shaped TENG with different surfaces, including **a** flat PET film: (a_1) 2 cm × 2 cm or (a_2) 2 cm × 4 cm; **b** flat PDMS film; **c** surface-nanostructured PDMS film; **d** PDMS film with groove-shaped microstructures; **e** PDMS film with groove-shaped micro-/nanohierarchical structures; **f** PDMS film with pyramid-shaped microstructures; **g** PDMS film with pyramid-shaped micro-/nanohierarchical structures

microstructures (Fig. 5.11d, f) and nanostructures (Fig. 5.11c). Compared with the performance of the flat PDMS film (Fig. 5.11b), the output voltage and the current of micro-/nanohierarchical structure with V-shaped grooves (Fig. 5.11e) are enlarged by 61.4 and 118 %, respectively, while the TENG with inverted pyramids (Fig. 5.11g) shows more attractive property in increasing the output voltage and current by 100 and 157 %, respectively, compared with the TENG with flat PDMS film. Although the micro-/nanohierarchical structures can further enhance the output performance of TENG compared to pure microstructures, such an enhancement is very limited. And therefore, microscale morphology plays a dominant role in improving the output performance of TENG.

Fig. 5.12 SEM images of silicon molds for fabricating PDMS films shown in Fig. 5.11

5.2.2 Enhancement of TENG Based on Hybrid Mechanisms

As described above, TENG has excellent characteristics of high output voltage and current, but the electric output is typically a pulse signal with short duration time. Therefore, from the point view of output electric power, the power density of TENG is still unable to meet the requirements of practical applications. We

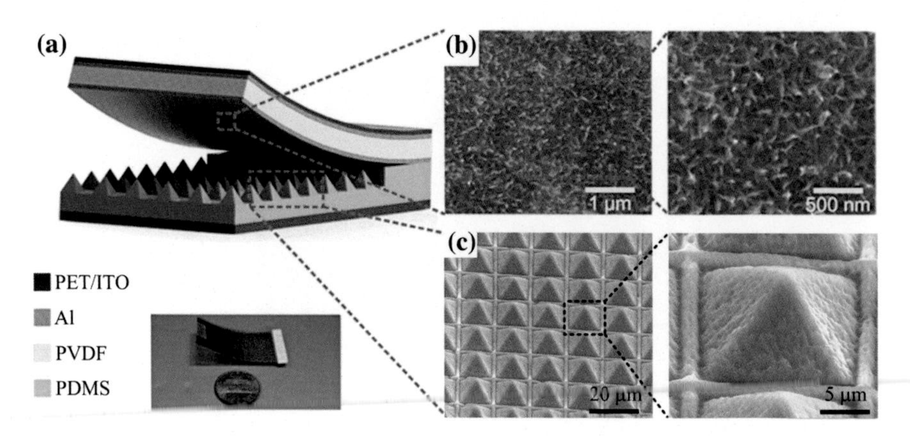

Fig. 5.13 **a** Schematic view and photograph of the r-shaped hybrid NG. **b** SEM images of the surface-nanostructured aluminum electrode. **c** The PDMS surface with pyramid-shaped micro-/nanohierarchical structures

proposed and realized two methods to enhance the power density of TENG in the above two sections. In this subsection, we propose another approach based on hybrid energy harvesting mechanisms to enhance the performance of nanogenerator (NG), i.e., integration of piezoelectricity and triboelectricity [14].

An r-shaped hybrid NG was designed by the combination of piezoelectric part and triboelectric part together, as shown in Fig. 5.13. This device consists of two parts, the upper PVDF–PET/ITO layer (i.e., piezoelectric NG) and the lower PDMS/Al layer (triboelectric NG). The two parts are fixed at one end, forming an r-shaped geometry. Under mechanical forces, the r-shaped hybrid NG can generate piezoelectric power as well as triboelectric power.

In order to enhance the output performance, we fabricated micro-/nanohierarchical structures onto the surface of PDMS film by micro-/nanointegrated fabrication technology presented above in this thesis and fabricated nanostructures onto the aluminum electrode of PVDF film by oxidation process using high-temperature DI water. Introducing micro-/nanostructures into triboelectric surfaces enlarges the effective triboelectric area and then enhances the output performance of NG.

Figure 5.14 illustrates the working principle of r-shaped hybrid NG. When the external force is applied (i.e., pressing step), the r-shaped PVDF film changes its profile from bent state to flat state. When the external force is released (i.e., releasing step), the PVDF film recovers to be r-shaped. Generally, during one cycle of the external force, the bottom Al electrode of PVDF and the PDMS film contact separates, and one triboelectric power signal is obtained. In the meantime, two opposite deformation procedures of PVDF film occur, and two piezoelectric signals are generated. Therefore, one triboelectric effect and two piezoelectric effects occur in one working cycle of this r-shaped hybrid device.

The output of this hybrid device can be divided into two parts, including triboelectric part and piezoelectric part, which were tested by connecting with a full-wave rectifier bridge. For the piezoelectric part, the output voltage achieved

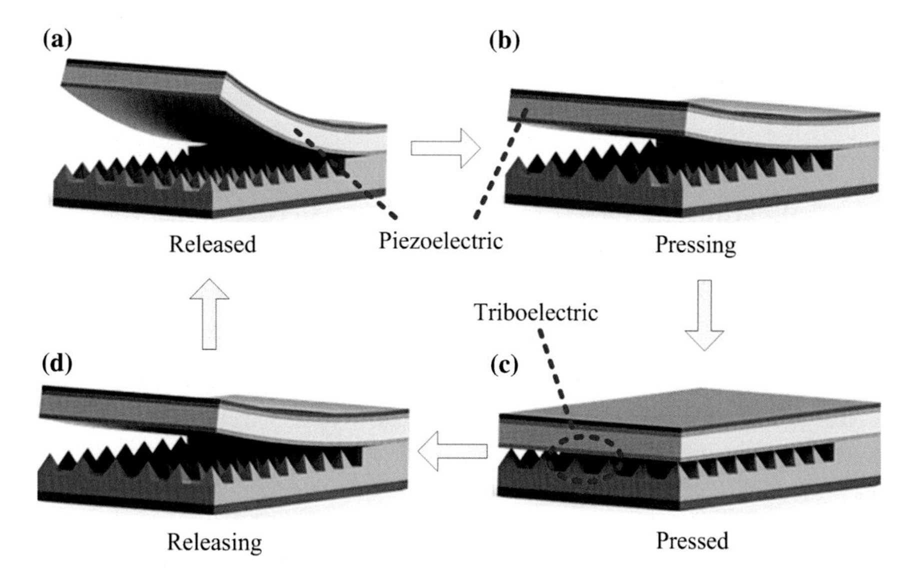

Fig. 5.14 Working principle of the r-shaped hybrid NG

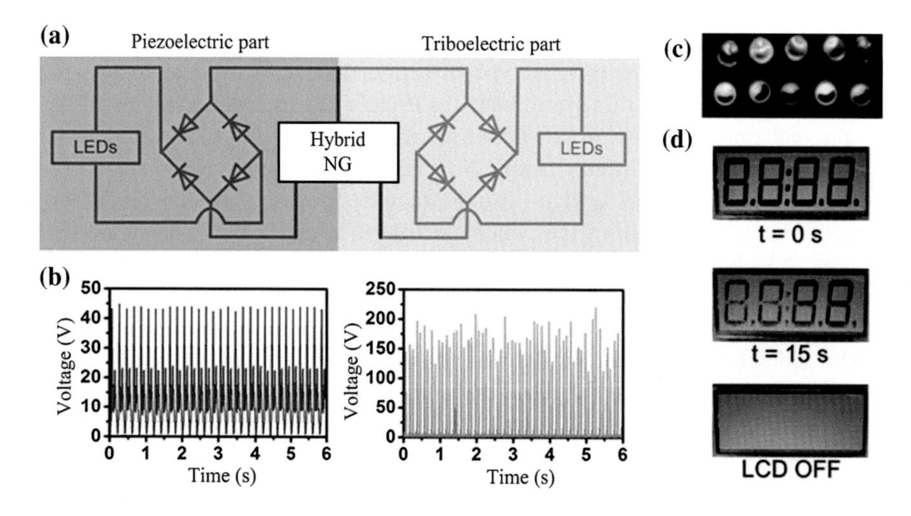

Fig. 5.15 Tests and measurement results of the output performance of r-shaped hybrid NG

to 45 V, while 200 V was obtained by the triboelectric part. Compared with the pure piezoelectric nanogenerator or triboelectric nanogenerator, this hybrid nanogenerator can be used to harvest the environmental energy simultaneously based on both of them and then enhance the harvesting efficiency. In addition, the fabricated hybrid device was demonstrated to power light-emitting diodes (LEDs) and a liquid crystal display (LCD), as shown in Fig. 5.15. Ten LEDs were directly lightened, while letters appear on the LCD screen for 15 s. In summary, this hybrid nanogenerator shows a remarkable ability of powering commercial electronics.

5.3 Applications of TENG for Biomedical Microsystems

In former sections of this chapter, we introduce the work of enhancing output performance of TENG, i.e., fundamental theory analysis, surface structural optimization, and single-step plasma treatment. Based on the above work, the output performance of TENG has been significantly improved. These high-performance TENGs are flexible, environmental-friendly, and robust, which show attractive potential applications for commercial electronics and biomedical microsystems, especially implantable electronic microdevices. This subsection is to explore the feasibility of applying these TENGs in the biomedical microsystems.

5.3.1 Application of TENG for Driving Neural Prosthesis In Vitro

The neutral prosthesis refers to a type of artificial devices used to recover or replace the impaired neutral tissues, which is usually of excellent biocompatibility and can realize the function of one or several kinds of neutral systems. Taking the retina neutral prosthesis as an example (see Fig. 5.16), it contains four parts: (1) nanogenerator, (2) control system (i.e., the module for receiving and processing signals), (3) neutral prosthesis (i.e., microelectrode arrays), and (4) sensing system (i.e., the module for capturing images and transmitting signals). The sensing system is used to capture the environmental images, which are transformed as electrical signals. Subsequently, these electrical signals are transmitted to the control system and processed there. Finally, the treated signals are utilized to stimulate the retinal nerves via the neural prosthesis, and then, the general images are reproduced in the brain to make blind people "seeing" the world. As for neutral prosthesis, the energy supply is always a matter of concern. Here, the sandwich-shaped TENG is utilized to directly drive an implantable 3-D microelectrode array for neutral prosthesis, thus showing the application potential of TENG in the biomedical microsystems. Figure 5.17 shows the test system of high-performance TENG driving the implantable 3-D microelectrode array in the phosphate-buffered saline (PBS), which measures the loop current by using a low-value resistance.

Figure 5.16(i), (ii) shows SEM images of the implantable 3-D microelectrode array, which is 3-D silicon-tip structure with 20 tips. The silicon-tip array was fabricated by the combination of anisotropic and isotropic etching, and then, it was transferred to a flexible and biocompatible substrate of parylene-C film. The patterned platinum electrodes with an intermediate layer of gold were realized via an

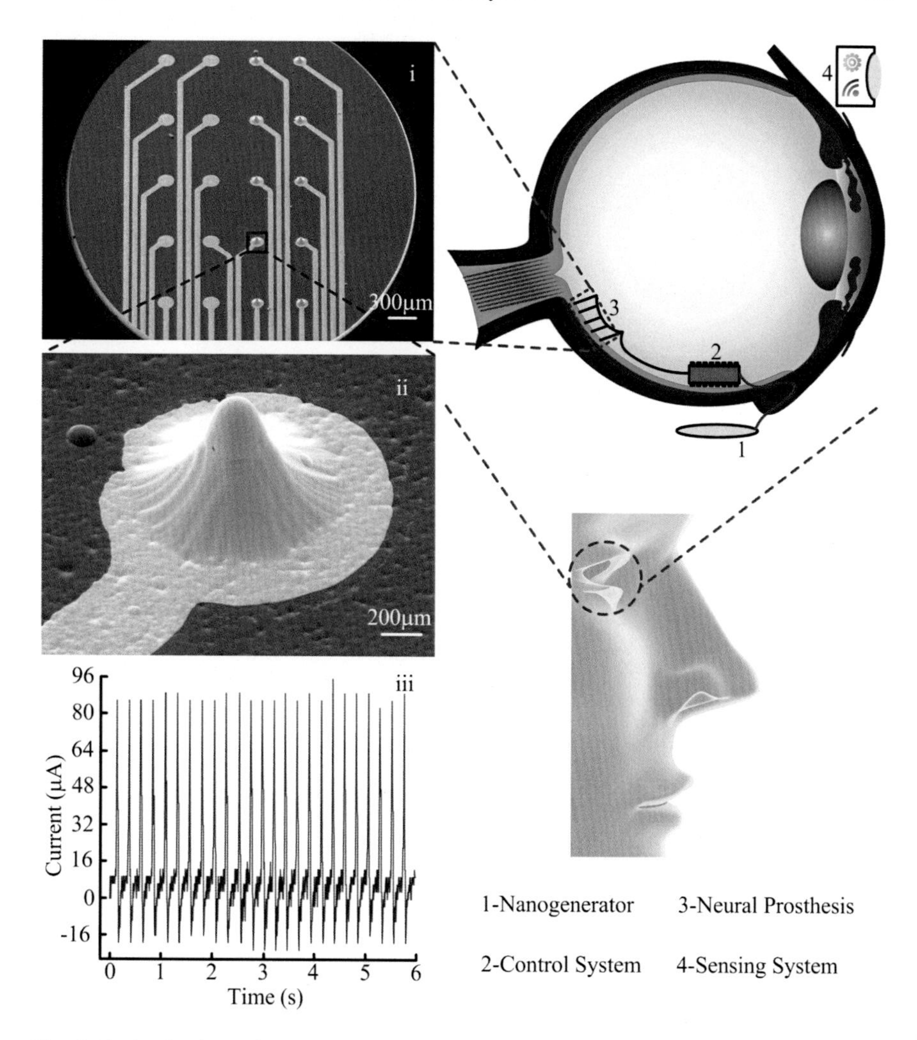

Fig. 5.16 Applications of the sandwich-shaped TENG for biomedical microsystem. (**i, ii**) Implantable 3-D microelectrode array for neural prosthesis was directly driven by TENG, and (**iii**) the current reached 88 μA

aluminum-photoresist dual-layer lift-off process. The detailed fabrication process of the above implantable 3-D microelectrode array was described in Refs. [15, 16]. To better reflect the working status of real implantation, we put the 3-D microelectrode array in the phosphate-buffered saline (PBS) solution, which is made of 4.0 g NaCl, 0.1 g KCl, 0.12 g KH_2PO_4, 1.82 g Na_2HPO_4, and 1000 mL DI water, and used as the artificial physiological environment for the test.

According to Fig. 5.16(iii), the measured current achieved to 88 μA, showing that the implantable 3-D microelectrode array was successfully driven. At this stage, there are still several challenges, such as flexibility, compatibility, and integration, to be overcome. However, this novel generator has shown an attractive

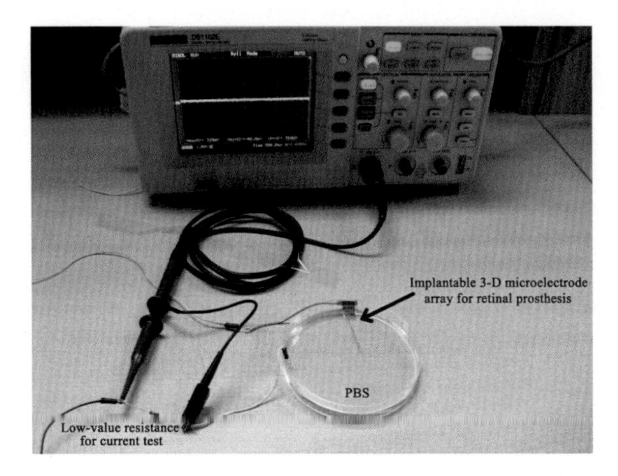

Fig. 5.17 Test system of high-performance TENG driving an implantable 3-D microelectrode array in PBS solution

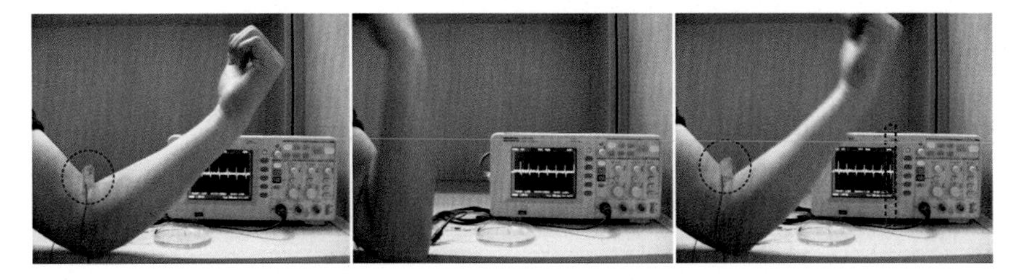

Fig. 5.18 Applications of the sandwich-shaped TENG to harvest mechanical energy from human arm movements and drive the implantable 3-D microelectrode array

application future, e.g., it can be placed on human arms, under feet, or even between two eyelids [17] to harvest the low-frequency mechanical energy and convert it to be frequency-multiplication high-output electric energy, which can be used to directly power biomedical micro-/nanosystems, as shown in Fig. 5.18.

5.3.2 Application of TENG for Driving Neural Prosthesis In Vivo

The transdermal stimulation electrode is very important for in vivo electroporation and repairing the impaired neutral network. To realize the function of stimulation, a simple electrode with double needles has been developed [18]. However, this double-needle electrode cannot provide uniform electrical field distribution and may damage the tissue due to the local ultra-high electrical field, while the 3-D microneedle electrode array (MEA) produces more uniform electric field and reduces the tissue damage.

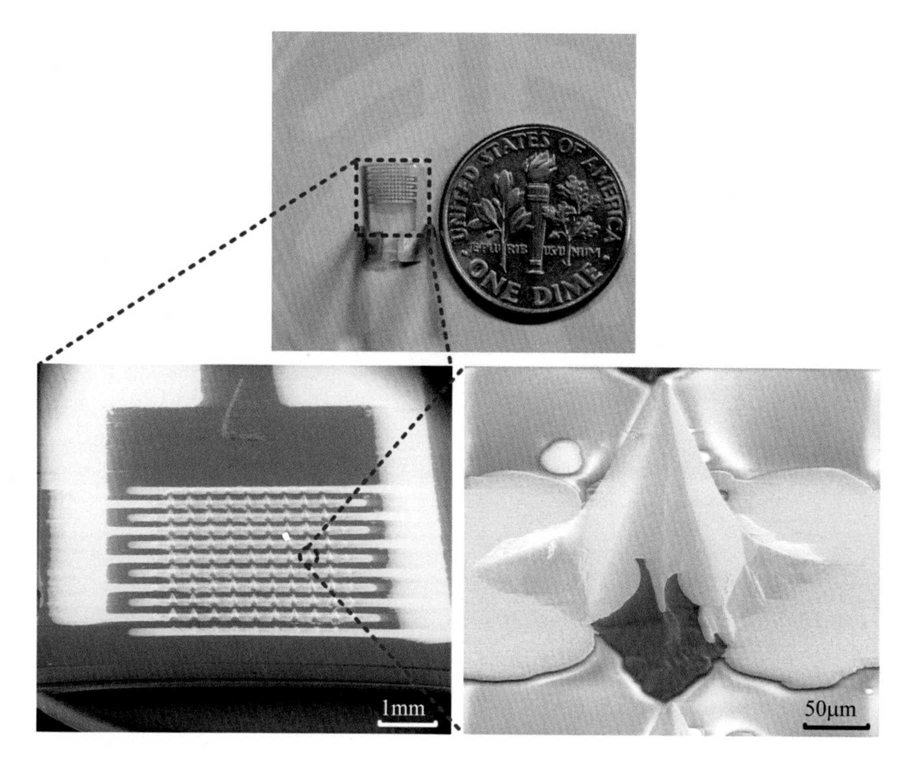

Fig. 5.19 The photograph of the flexible implantable microneedle electrode array and SEM images of the microneedle electrode array

Figure 5.19 shows the photograph of the flexible implantable microneedle electrode array and SEM images of the microneedle electrode array. As the height of the MEA reaches 190 μm, it is also called as 3-D transdermal electrode array [19]. To achieve better electrical stimulation effects, this MEA is designed to comb-shaped geometry. This MEA consists of 9 × 9 Si-based high-aspect-ratio tips, and the electrode connection ports are derived from two ends. This design can result in a uniform and stable stimulation electric field in the electrode array area under an external electric field.

In our experiments, this MEA was implanted into a real frog tissue such that the high-aspect-ratio tips tightly contacted with the frog's sciatic nerve, as shown in Figs. 5.20 and 5.21. To ensure the activity of frog's muscle tissue and neural system, the frog tissue was placed in an amphibian saline solution (i.e., Ringer's solution) so that the experimental results are closer to the nerve tissue electrical stimulation of live animals. The Ringer's solution is made of 6.6 g NaCl, 0.15 g KCl, 0.15 g $CaCl_2$, 0.2 g $NaHCO_3$, and 1000 mL DI water.

Figure 5.21 illustrates the real-time response of frog's leg to the stimulation of the microneedle electrode array driven by TENG. As shown in Fig. 5.21(i), three parts of frog leg curled up together in the initial state. When using finger to press

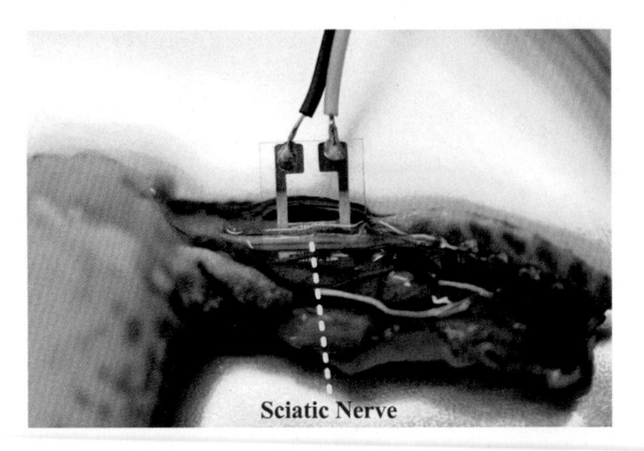

Fig. 5.20 The photograph of the nerve stimulation test system, in which TENG was used to stimulate the frog's sciatic nerve via the implanted MEA

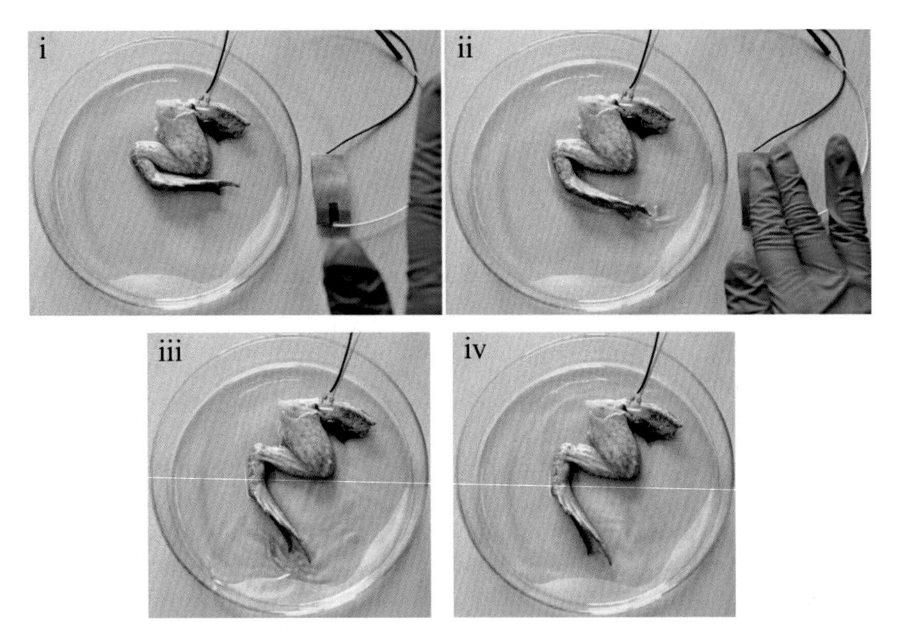

Fig. 5.21 Sequential photographs of the real-time response of frog's leg by the stimulation of the microneedle electrode array powered by TENG

the TENG, the instantaneous output voltage of the TENG induces the loop current among MEA via the sciatic nerve. Therefore, the sciatic nerve is stimulated by the loop current and actuated the leg muscle of frog, as shown in Fig. 5.21(ii–iv). The above electrical stimulation response is not an accidental phenomenon, and its reliability has verified by continuous tapping TENG to induce the movement of frog's leg. The details can be referred to the experimental video in Supporting Information in the author's published paper [20].

5.4 Applications of TENG in the Other Fields

5.4.1 Self-Powered Humidity Monitoring Sensor

As a power generation approach harvesting the ambient energy, the influence of environment itself on the performance of TENG cannot be ignored. Therefore, we will discuss the effect of environmental factors on TENG performance in this subsection. Humidity is one of the most important factors affecting TENG performance in practical situations [21, 22], due to that too many water molecules will accelerate the dissipation of triboelectric charges. According to Fig. 5.22, the TENG output voltage decreases continuously as the relative humidity increases from 10 to 90 %. To more accurately illustrate this change process, we define the normalized voltage as shown in Eq. (5.2),

$$V_{\mathrm{N}} = \frac{V}{V_0} \tag{5.2}$$

where V_0 is the output voltage with the relative humidity of 10 %.

According to Fig. 5.22, the normalized voltage dramatically decreases by 43 % as the relative humidity increases from 10 to 50 %, while the normalized voltage is reduced by 23 % when the relative humidity increases from 50 to 90 %. That is to say, the decrease rate of the output voltage becomes gradually small as the relative humidity increases. Therefore, we proposed and realized a new-type self-powered humidity

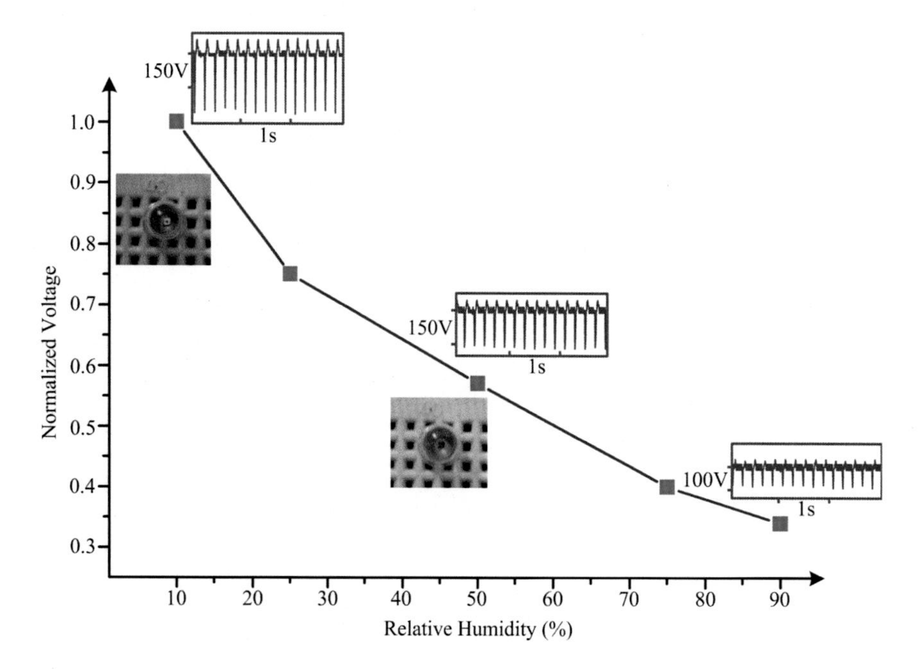

Fig. 5.22 The relation between environmental humidity and TENG output voltage

monitoring sensor, as is shown in Fig. 5.22. A commercial light-emitting diode (LED) lamp with a matching variable resistor is used as the indicator to replace the expensive electrometer to monitor the change of environmental humidity, which shows the possibility to simplify the detection system. When the relative humidity is lower than 50 %, the LED lamp is turned on. Oppositely, the LED lamp is turned off when the relative humidity is larger than 50 %. Thus, the above simple device can semiquantitatively indicate the humidity in the range of larger than 50 or below 50 %.

5.4.2 Application for Portable Electronics

The r-shaped hybrid NG has attractive advantages in the practical applications, and its r-shaped structure facilitates the effective harvesting of various kinetic energies in the environment. For example, integrating it at the bottom of shoes can harvest the kinetic energy during human walking or running. Besides, we can also integrate it in various instruments or devices with moving parts and harvest the energy during their movements. We successfully realize an integrated keyboard system to harvesting the daily use energy, e.g., the mechanical energy produced by tapping the keyboard, as shown in Fig. 5.23.

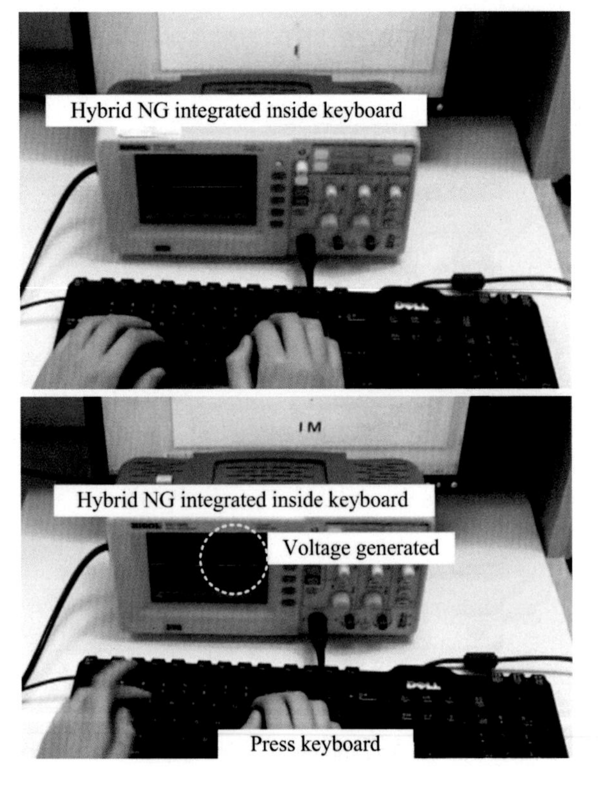

Fig. 5.23 The demonstration of integrating an r-shaped hybrid nanogenerator under a keyboard

Here given a simple presentation, we integrate the r-shaped hybrid NG under the spacebar of computer keyboards. Actually, due to the advantage of miniaturization of hybrid NG, it can be integrated under the whole keyboards. Therefore, when using the keyboard and hitting the space key for work, the spacebar will move down in the external force, thereby exerting downward pressure on the hybrid NG at the bottom to cause bending deformation. After the finger departs away from the space key (i.e., release external force), the spacebar will return to the initial position, and the hybrid NG will also recover to the initial position. The above process is consistent with the working principle of the hybrid NG in Fig. 5.14 and thus realizes the effective conversion and harvesting of mechanical energy produced by using the keyboard. Figure 5.23 presents the real-time measurement results by using the digital oscilloscope. In summary, the r-shaped NG owns an excellent characteristic of easy integration and has widespread applications in practical situations.

5.5 Conclusions

In this chapter, we first summarize theoretical studies of TENG and then perform theoretical calculations on the vertical ionization energies of triboelectric materials at the molecular level based on the density functional theory to analyze the ability of capturing or losing electrons and explore the nature of triboelectrification effect.

On this basis, we propose three methods to enhance the TENG output performance. The first one is modifying the surface of the friction material based on single-step fluorocarbon plasma treatment. The second one is designing and optimizing the surface structures of friction at the micro-/nanolevel. The third one is piezoelectric and triboelectric hybrid NG.

First, this chapter investigates the single-step fluorocarbon plasma treatment in depth, explored the influence of key process parameters on the device output performance, reveals the two opposite effects (i.e., chemical enhancement and physical diminution) of the fluorocarbon plasma treatment on the surface of the friction material, and finally obtains the optimal process parameters. In the meantime, to make this method stable, reliable, and general, we carry out in-depth investigation and analysis on it in terms of processing equipment, plasma gas, substrate material, continuous working, and long-term test. Comparative experiment results show that this single-step fluorocarbon plasma treatment process is simple and not limited to substrate materials and processing equipment and that the output performance of modified TENG is stable and reliable. Therefore, this new method is single step, robust, and universal for enhancing the output performance of TENG.

Second, this chapter systematically studys the relationship between output performance of TENG and the surface micro-/nanostructures of friction layers, demonstrating that the surface roughness increase can enhance TENG's performance by comparisons among several kinds of surface morphologies including smooth surface, nanoscale structures, microscale structures, and micro-/nanohierarchical structures, and further points out that microscale structures play a leading role in enhancing the TENG performance.

Subsequently, a r-shaped hybrid nanogenerator is proposed, which effectively integrates the triboelectric effect and the piezoelectric effect and enhances the output performance due to simultaneously harvesting both of triboelectric and piezoelectric energies. In addition, the powering ability of this hybrid nanogenerator is investigated by systematical electric test and lightening up LED and LCD.

Finally, the applications of the presented high-performance flexible TENG are explored, including constructing self-powered sensors and powering commercial electronics and biomedical microsystems. We firstly realize a self-powered humidity monitoring sensor and an integrated keyboard system harvesting the mechanical energy. Besides, the frog's sciatic nerve is successfully stimulated by an implantable 3-D microneedle electrode array driven by TENG. This is the first demonstration of nanogenerator successfully powering an implantable biomedical microsystem to stimulate the real biological tissue, which promotes a major step of practical applications of TENG.

References

1. X.S. Zhang, M.D. Han, R.X. Wang, B. Meng, F.Y. Zhu, X.M. Sun, W. Hu, W. Wang, Z.H. Li, H.X. Zhang, High-performance triboelectric nanogenerator with enhanced energy density based on single-step fluorocarbon plasma treatment. Nano Energy **4**, 123–131 (2014)
2. S. Wang, L. Lin, Z.L. Wang, Nanoscale triboelectric-effect-enabled energy conversion for sustainably powering portable electronics. Nano Lett. **12**, 6339–6346 (2012)
3. Z.L. Wang, Triboelectric nanogenerators as new energy technology for self-powered systems and as active mechanical and chemical sensors. ACS Nano **7**, 9533–9557 (2013)
4. S. Niu, S. Wang, L. Lin, Y. Liu, Y.S. Zhou, Y. Hu, Z.L. Wang, Theoretical study of contact-mode triboelectric nanogenerators as an effective power source. Energy Environ. Sci. **6**, 3576–3583 (2013)
5. S. Niu, Y. Liu, S. Wang, L. Lin, Y.S. Zhou, Y. Hu, Z.L. Wang, Theory of sliding-mode triboelectric nanogenerators. Adv. Mater. **25**, 6184–6193 (2013)
6. J.P. Perdew, K. Burke, M. Ernzerhof, Generalized gradient approximation made simple. Phys. Rev. Lett. **77**, 3865–3868 (1996)
7. ADF 2010.01, http://www.scm.com
8. C.F. Guerra, J.G. Snijders, G. te Velde, E.J. Baerends, Towards an order-N DFT method. Theoret. Chem. Acc. **99**, 391–403 (1998)
9. G.T. Velde, F.M. Bickelhaupt, E.J. Baerends, C.F. Guerra, S.J.A. van Gisbergen, J.G. Snijders, T. Ziegler, Chemistry with ADF. J. Comput. Chem. **22**, 931–967 (2001)
10. E. van Lenthe, E.J. Baerends, Optimized Slater-type basis sets for the elements. J. Comput. Chem. **24**, 1142–1156 (2003)
11. C. Lee, W. Yang, R.G. Parr, Development of the Colic-Salvetti correlation-energy formula into a functional of the electron density. Phys. Rev. B **37**, 785–789 (1988)
12. A.H. Kuptsov, G.N. Zhizhin, *Handbook of Fourier transform Raman and infrared spectra of polymers*, 1st edn. (Elsevier Science, Amsterdam, 1998)
13. G. Socrates, *Infrared characteristic group frequencies*, 2nd edn. (Wiley, New York, 1994)
14. M.D. Han, X.S. Zhang, B. Meng, W. Liu, W. Tang, X.M. Sun, H.X. Zhang, r-Shaped hybrid nanogenerator with enhanced piezoelectricity. ACS Nano **7**, 8554–8560 (2013)

15. R. Wang, W. Zhao, W. Wang, Z. Li, A flexible microneedle electrode array with solid silicon needles. J. Microelectromech. Syst. **21**, 1084–1089 (2012)
16. R. Wang, X. Huang, G. Liu, W. Wang, F. Dong, Z. Li, Fabrication and characterization of a Parylene-based three-dimensional microelectrode array for use in retinal prosthesis. J. Microelectromech. Syst. **19**, 367–374 (2010)
17. S. Lee, S.H. Bae, L. Lin, Y. Yang, C. Park, S.W. Kim, S.N. Cha, H. Kim, Y.J. Park, Z.L. Wang, Super-flexible nanogenerator for energy harvesting from gentle wind and as an active deformation sensor. Adv. Funct. Mater. **23**, 2445–2449 (2013)
18. H. Aihara, J. Miyazaki, Gene transfer into muscle by electroporation in vivo. Nat. Biotechnol. **16**, 867–870 (1998)
19. R. Wang, Z. Wei, W. Wang, Z. Li, in *Flexible microneedle electrode array based-on Parylene substrate*. The 16th International Conference on Miniaturized Systems for Chemistry and Life Sciences (Okinawa, Japan, 28 Oct–1 Nov 2012), pp. 1249–1251
20. X.S. Zhang, M.D. Han, R.X. Wang, F.Y. Zhu, Z.H. Li, W. Wang, H.X. Zhang, Frequency-multiplication high-output triboelectric nanogenerator for sustainably powering biomedical microsystems. Nano Lett. **13**, 1168–1172 (2013)
21. V. Nguten, R. Yang, Effect of humidity and pressure on the triboelectric nanogenerator. Nano Energy **2**, 604–608 (2013)
22. E. Németha, V. Albrechtb, G. Schubert, F. Simon, Polymer tribo-electric charging: dependence on thermodynamic surface properties and relative humidity. J. Electrostat. **58**, 3–16 (2003)

Chapter 6
Summary and Perspectives

6.1 Summary

The development of nanofabrication technology is currently restricted by the minimum lithography scale and the difficulty for achieving high-throughput manufacturing. Regarding the above issues, this thesis presents a systematic approach to fabricate large-scale micro-/nanohierarchical structures based on traditional microfabrication techniques. The presented micro-/nanointegrated fabrication technology has been demonstrated to be simple, reliable, compatible with the standard CMOS process, applicable for mass production, and cost-effective, which can be used to manufacture micro-/nanohierarchical structures onto silicon-based materials and flexible materials. The fabricated samples show several attractive properties, such as super-hydrophobicity, wideband anti-reflectance, and surface-enhanced Raman scattering. In the meantime, the flexible micro-/nanohierarchical materials fabricated by the as-presented fabrication technology were successfully applied in fabricating high-performance flexible triboelectric nanogenerators, which extends the application of micro-/nanohierarchical material in microenergy field.

Briefly, the research work in this thesis can be summarized as follows (as shown in Fig. 6.1):

Firstly, this thesis presents and realizes a systematic and large-scale micro-/nanointegrated fabrication technology which can be applied for fabricating micro-/nanohierarchical structures onto both Si-based substrate and flexible materials. This fabrication technology has various merits including simple operation, good reliability, controllability and universality, and mass-production ability.

Secondly, this thesis systematically explores the core scientific issue of micro-/nanointegrated fabrication and intensively studies the multiscale coupling effect and the size effect to reveal their underlying mechanism through experimental tests, theoretical analysis, performance characterization, etc.

X.-S. Zhang, *Micro/Nano Integrated Fabrication Technology and Its Applications in Microenergy Harvesting*, Springer Theses, DOI 10.1007/978-3-662-48816-4_6

Fig. 6.1 Summary of the research work in this thesis

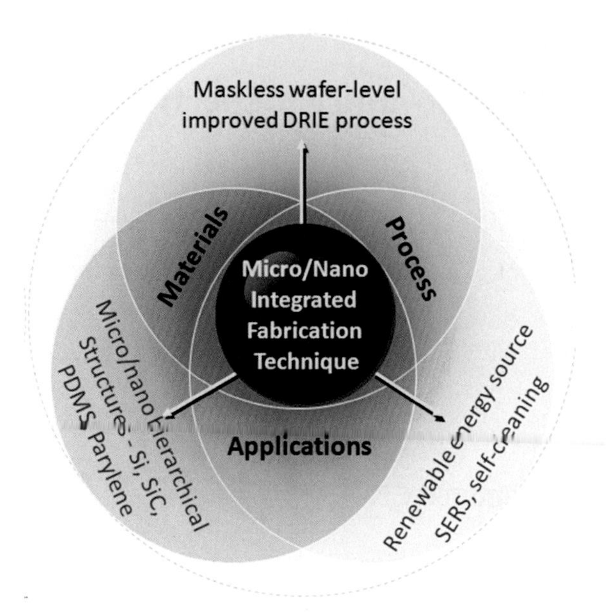

Finally, this thesis further introduces the as-presented micro-/nanointegrated fabrication technology to the microenergy field. Several high-performance nanogenerators are proposed and demonstrated to be a robust micropower source. In the meantime, their operating mechanism, theoretical foundation, electrical performance, reliability and stability, and applications in the field of biomedical microsystems are intensively analyzed and researched.

6.2 Main Contribution

This thesis carries out extensive research about large-scale micro-/nanointegrated fabrication technology based on traditional microfabrication techniques and its applications in microenergy field. It has significant innovations in theory, methods, and applications. The main innovations are summarized as follows:

Firstly, this thesis proposes and realizes a large-scale micro-/nanointegrated fabrication technology. This fabrication approach has advantages including simple process steps, large-scale mass production, low cost, etc., and it can be widely applied on the fabrication of micro-/nanohierarchical structures onto Si-based substrate and flexible materials. The optical properties, liquid dynamics, and spectroscopic characteristics were also studied systematically. The fabricated micro-/nanohierarchical structures show excellent characteristics such as super-hydrophobicity, wideband anti-reflectance, and surface-enhanced Raman scattering.

Secondly, this thesis systematically studies the multiscale interaction effect during micro-/nanointegrated fabrication procedure. The above new phenomena observed from silicon and flexible materials were intensively analyzed by using vector field analysis and finite element simulation, respectively. The underlying mechanism of the interaction between microstructures and nanostructure is explored and revealed.

Thirdly, we employed the density functional theory into the analysis of the hydrophobic polymer material characterization and the triboelectric nanogenerator to resolve the underlying mechanisms of hydrophilic and hydrophobic properties of the polymer material in the molecular and atomic level as well as the essential factors of capturing or losing electrons during the electrification effect.

Finally, several high-performance nanogenerators were proposed, and we constructed the theoretical and physical model of triboelectric nanogenerator and gave its numerical analysis. Several universal methods were developed for improving the output performance of triboelectric nanogenerator. In addition, the fabricated high-performance flexible triboelectric nanogenerators were demonstrated to powering biomedical microsystems and commercial electronics.

6.3 Perspectives

In summary, this thesis proposes a universal micro-/nanointegrated fabrication technology. The proposed approach can be applied to the fabrication of a variety of common materials, which has been confirmed in crystalline silicon, PDMS, and parylene-C. According to the analysis in Chap. 2, it is clear that the improved DRIE process, which is the core of the presented micro-/nanointegrated fabrication technology, is not sensitive to the fabrication material. Therefore, it can be applied to various materials in principle. However, we are unable to try more materials regarding the limited time available. Nevertheless, we still carried out some preliminary work based on silicon carbide and successfully realized surface-nanostructured SiC thin films. By using the improved DRIE process presented here, we fabricated high-dense nanopillar forest on the SiC film grown by plasma-enhanced chemical vapor deposition (PECVD) process on silicon substrate. After the removal of silicon substrate by using KOH wet-etching process, the flexible SiC film with high-dense nanopillar forest on the surface was obtained. The fabricated SiC samples show a tunable wettability that changes from super-hydrophobicity to super-hydrophilicity, and an attractive property of surface-enhanced Raman scattering with an enhancement factor of 3.4×10^4 was also observed. More details about this work can be found in our published paper [1].

As a prospect from the successful attempt above, it is believed that when this technique is further combined with more other microprocessing techniques, we can successfully achieve the preparation of SiC micro-/nanohierarchical materials. Therefore, in the future, we can use it in many other materials to carry out research work and explore more new micro-/nanohierarchical materials with many other excellent properties.

This thesis also explores the wide applications of micro-/nanohierarchical structures in microenergy field, proposes a variety of high-performance flexible triboelectric generators, and demonstrates their applications in commercial electronics and biomedical microsystems. Thus, it is expected that these flexible triboelectric generators based on the proposed micro-/nanointegrated fabrication technology have attractive prospect in the field of microenergy source because of its advantages of high-output performance, low cost, and easy integration. As one of the innovations of this work, the integration of flexible high-performance triboelectric generators and biomedical microsystems for practical implantable self-powered neural prosthesis is feasible and can be continued in the future.

In addition, according to the summary in Chap. 1, along with super-hydrophobic, super-hydrophilic, and anti-reflective properties, the micro-/nanohierarchical structures also show other remarkable properties, such as super-adhesion. The above features will thrive rapidly by employing flexible materials with micro-/nanohierarchical structures. Therefore, as an attractive future vision, researchers could study and explore the properties which have been exhibited as well as the properties which have not been found, and further extend its widespread use in many fields. Finally, this thesis realizes totally four kinds of micro-/nanohierarchical structures, including the pyramid shape and the groove shape with cross-sectional views of inverted triangular and trapezoid. In the future, a variety of different micro-/nanohierarchical structures with different sizes can be investigated to achieve the optimization and innovation of materials' properties and functions.

Reference

1. X.S. Zhang, B. Meng, F.Y. Zhu, W. Tang, H.X. Zhang, Switchable wetting & flexible SiC thin film with nanostructures for microfluidic surface-enhanced Raman scattering sensors. Sens. Actuators A **208**, 166–173 (2014)

Printed in the United States
By Bookmasters